COMMUNICATING

RELIABILITY

RISK &

RESILIENCY

to DECISION MAKERS

"HOW TO GET YOUR BOSS'S BOSS TO UNDERSTAND"

JD SOLOMON

Table of Contents

INTRODUCTION   i

CHAPTER 1: RELIABILITY, RISK, AND RESILIENCY   1

CHAPTER 2: OTHER CONCEPTS THAT CONFUSE   17

CHAPTER 3: GRAPHICAL EXCELLENCE   35

CHAPTER 4: THE ROLE OF ETHICS   53

CHAPTER 5: PRACTICAL TOOLS   67

CHAPTER 6: NOISE   89

CHAPTER 7: THE AUDIENCE   111

CHAPTER 8: RESILIENCY AND RARE EVENTS   139

CHAPTER 9: PLANS FOR COMMUNICATIONS   157

NOTES   181

GLOSSARY   193

BIBLIOGRAPHY   195

INDEX   203

ACKNOWLEDGEMENTS

Tim Adams and Dan Vallero provided the primary technical reviews. I was very fortunate to have both spend quality time helping me to sharpen the technical discussions and avoid fatal flaws. I am also appreciative of Tim's helpful writing tips and detailed proofreading.

Jennifer Haynie served as my editor for this book. An accomplished fiction writer, Jennifer has an undergraduate degree in physics and a graduate degree in environmental management. That combination was truly beneficial in my effort to make this readable for the practitioner and to preserve the first-person nature of the way I present this information in a live setting. I am also appreciative of her patience, suggestions, and coaching in helping me get this book published.

Sean McLean provided the illustrations. It was our first project together and his first venture into non-fiction. He needed some patience on this one. His extra effort is greatly appreciated.

I am very appreciative to my beta-readers who dug into the initial draft and told me what they thought. It is a great and trusted group: my son and aerospace engineer, Kyle Solomon; my friend, attorney, and former race team manager, Andy Anderson; my engineering colleague, Chris Ericksen; and my father, John Solomon, who spent his career working with a wide range of technical specialists designing some big and ground-breaking power generation facilities. My budding attorney and daughter, Kelley Solomon, provided key review of the final document.

I have thought it a bit trite for a writer to thank their spouse in the acknowledgements. But now I understand why they do. For months, the spouse gets to hear erratic snippets and with little notice is asked to provide meaningful input on a small matter with little context. At least that is how it works in my home. Many thanks to my wife, Kristine, for her input and her patience.

Finally, many thanks to the participants in my webinars, workshops, and guest lectures for their feedback along the way. During the writing of this manuscript, I have reviewed some of those old notes and even made a few follow-up calls and emails. I am very appreciative of those scattered debates, questions, and comments which have collectively made me better. And it has made this book better.

FORWARD

One common thread that runs through every technical discipline is that the principal client is the public. Every professional code of ethics includes language that places a burden on the professional for protecting the public. In particular, the engineering code of ethics establishes a high bar by requiring that the engineer must "hold paramount the safety, health and welfare of the public." In this book, the author has substantially aided the technical professional in the quest to solve complex problems by helping to communicate and translate ethical canons for the decision maker, who may or may not be an engineer or who may have only a rudimentary understanding or risk-based decision making.

The technical professional must balance sound science with feasible solutions by placing reliability and risk within the context of agile and resilient solutions to problems related to the built and natural environment. In a very understandable way, J.D. Solomon has used cases and examples to explain the statistical, technical and ethical demands on the dynamics of decision making. J.D. brings to life the complex nature of engineering decision making. This book is not simply a "once-read". It is a keeper that should be consulted before the next important briefing of friendly or hostile managers, customers, clients, partners, or colleagues around the proverbial water cooler.

I have had the distinct pleasure of working with the author, especially in presenting at engineering workshops and conferences. In doing so, he has shared so many insightful ideas that it has been difficult to keep track of them. In particular, I have found myself envious of his ability to bring my often theoretical and academic understanding of reliability, risk and resilience to life within the real world. Therefore, I am gratified that he has been able to capture some of this wisdom in this book.

Also, I strongly encourage the reader to combine the knowledge here with one of J.D.'s workshops, where he puts these concepts into action. The best way to learn is by doing, but reading combined with the workshops is good preparation for the "doing".

Sometimes there is no precedent or clear way to know what to do. This is particularly challenging because doing is not possible without at least a modicum of understanding and agreement with decision makers. Further, these folks may fail to see how reliability and risk relate to their expectations. Notwithstanding the soundness of the technical recommendation, decision makers are not necessarily going to accept a recommendation that deviates from their expected costs or even if is different from an expected approach based on legal, financial, political and other perspectives.

Often, it is not the quality of the recommendation, but the manner in which it is conveyed and justified. Herein lie some very good ways to prevent miscommunications and, hopefully, reach agreement on the most reliable and lasting systems.

The Resilience Engineering Association ascribes successful engineering "to the ability ... to anticipate the changing shape of risk before damage occurs; failure is simply the temporary or permanent absence of that." I wish I had this book the last time I had to try to explain resiliency to my students.

Daniel A. Vallero, PhD
Duke University, Pratt School of Engineering

INTRODUCTION

The genesis of this book dates back about four years. In searching for interesting and relevant risk and reliability topics for papers and presentations, I asked numerous risk and reliability professionals what was the hardest part of their job. I expected responses related to a deeper understanding of some type of emerging technical topic or perhaps the better integration of technical topics. Instead, I got an almost universal reply that help was needed "to get senior management to understand what I do." When I questioned this answer, the reply again was almost universally, "I have the tools and knowledge to do the technical part; I just need to get them to understand."

This should not have been any surprise to me. After 30 years as an engineering consultant, the issue of technical professionals not being able to communicate effectively is a common one. The issue is often just as big for non-technical professionals and perhaps worse when over-confidence and lack of preparation intersect. I often tell technical audiences that if they were good at communicating with other humans, they would likely have had academic majors in sales, marketing, business, or political science. They majored in a technical profession because they were good at math and science but were not as proficient with human interaction. Some look at me as if I am not telling the truth. I ask the hypothetical question "what would your parents or siblings say if I asked them?" They uncomfortably chuckle. It helps deflect some of the truth's pain.

Our practice standards and our guidance manuals provide little help. My standard examples of this issue are the international risk standard, ISO 31000, and the international asset management standard, ISO 55000. ISO 31000 discusses communication as an important competency. In its depiction of a risk management framework, communication is interactive with five of the other six aspects of the framework. Yet discussion of the topic spans less than one column on a two-column page. ISO 55000 spends a little more space on the topic but does not provide anything except high level jargon.

I developed a short presentation that I used successfully for 30-minute conference sessions and keynote addresses. It proved to be a very popular topic. I combined this presentation with a training series I developed in the late 1990s and early 2000s. It then became the four-hour and eight-hour workshops that I reference in this book. These workshops proved to be equally successful and in demand as the short presentation. I was encouraged by some good friends who attended my events at the 2016 Society of Maintenance and Reliability Professionals (SMRP) Annual Conference and the 2016 Palisade National Risk Conference in New Orleans to convert my workshops notes into a book on the topic.

This book and the workshops are intended to reinforce each other and fill a gap in the industry. Both are meant to be mostly practical and less academic. Neither will fully address the depth of the topics of communication, reliability, risk, resiliency, and decision making—there are many other good references in the market on each of those topics. The gap this book seeks to fill is the practical integration of these topics in a way to make technical professionals more effective at what they do.

Chapters 1 through 4 provide foundational definitions and concepts. I often use the saying "a problem well defined is a problem half solved." It is equally true for communicating reliability, risk, and resiliency to decision makers.

Chapters 5 through 9 get into specific topical details and practical examples. Chapter 5 is wholly about the practical aspects and lessons learned. Chapter 6 explores how we communicate and successfully when dealing with different types of risk-based and reliability-based decisions. Chapter 7 explores the audience and practical considerations related to their impact on communications. Chapter 8 addresses resiliency and the special case of rare events. Chapter 9 brings everything together by providing practical frameworks and tips for developing the communications.

The origins of this book dates back 25 years. It was at that time that I started managing my first department of technical professionals who were required to collect data, analyze it, and communicate it effectively to internal and external stakeholders. This was about the same time that I became involved in community planning, economic development, and politics. In the mid-1990s, I developed my first series of training modules to facilitate making my technical staff better technical consultants. The eight-module series grew to a standard ten-module series by 2004. Four of those modules—presentation skills, quality calls, personality profiles, and communicating with the media and elected officials—have been key foundations for the "Communicating Reliability, Risk, and Resiliency to Decision Makers" workshops and this book.

Three unique individuals perhaps best bookend a wide range of people who have provided the span of technical input behind this book and the workshops. The late Colonel Roy G. Sowers and I worked together in the 1990s. Roy was an extraordinary figure by all measures. At one point in his career, he lost a narrow race for Lieutenant Governor in North Carolina and later became the state's first Secretary of the Department of Natural Resources and Economic Development. While in the service, he had served as the US Army's Chief of Psychological Warfare for the North American Armies. He also had some pretty good reference material stacked away in his personal archives. Roy knew and understood politics, business, and, most importantly, people. I was very fortunate to have him as a friend and mentor.

Dr. Daniel Vallero has provided much valuable input and collaboration over the past couple of years. Dan and I have written several papers together and co-presented on ethics at professional conferences. Also, he has allowed me to be a guest lecturer in his classes at Duke University. Dan has authored thirteen books on a full range of engineering topics including hazardous waste, air quality, water quality, disease, and, in most recent times, genetic engineering. We have had many meaningful technical discussions and debates on risk, ethics, and communications. His friendship and collaboration are greatly appreciated and have helped shape my thinking.

Tim Adams and I have enjoyed many hours of discussion and debate over the last couple of years. Tim has been at NASA for many years and is a true renaissance man who is equally deep in his knowledge of teaching, mathematics, reliability, managing for performance and communications. Like Roy, he understands people, how they process data and information, and how they behave under risk and uncertainty. Similar to Dan, Tim has a quick memory and is a walking encyclopedia of approaches and theories.

I have also learned much from additional formal training, my colleagues, service of state boards dealing with decisions related to technical information, and being on the front lines of business and public policy just about every day and certainly every week. My colleagues Adam Sharpe, Kathryn Benson, Jennifer Bell, and I have worked together for more than a decade communicating a wide range of reliability, risk, and resiliency issues to decision makers on some of their biggest, most complex, and career-changing projects and programs. We have had our successes and our failures, have learned much along the way, and have the battle scars to prove it. This topic is certainly one that requires collaboration and trusted advisors. I have certainly been fortunate to have those.

In closing these introductory remarks, I remind the reader once again that this book and the workshops on which it is based are simply a response to a practical gap in the industry. It is not intended to be a comprehensive discussion of all of the different aspects of communicating, reliability, risk, resiliency, and decision making. There are many other references in each of these subject areas. The book is intended to be a practitioner's reference. And whether you agree or disagree with some of the points, I hope that you will at least take away a few golden nuggets that will make you more effective in your practice.

CHAPTER 1

RELIABILITY, RISK, AND RESILIENCY

I usually start my workshops by having each attendee pair with another attendee. The exercise that follows is a simple one. I hand each pair of attendees two different envelopes. Each person opens an envelope and asks their partner the simple question that is written on the slip of paper inside. That simple question is either "What is Risk?" or "What is Reliability?"

The results are always entertaining. First, there are often about as many different answers to each question as there are participants in the workshop. Second, very few participants can do it quickly in one simple sentence. In fact, it usually takes so much time that I interrupt the explanations to keep things moving and get the end result of a simple, quick exercise.

This is probably not a surprising result when we consider the population as a whole. However, these are technical professionals who routinely communicate technical information, and usually some form of risk and reliability, to decision makers. And these are usually the upper tier of technical professionals who care enough to have signed up for a workshop about communicating risk and reliability. It is surprising that so few have a good, straightforward definition of risk and reliability prior to the workshop.

RELIABILITY

Reliability has the most universal definition from a technical perspective. It is most often defined as "the probability that an item will perform its intended function for a specified interval under stated conditions." This definition was somewhat institutionalized in the post-World War II era by the aerospace and US military sectors and has been more or less adopted as the dominate definition among all industry sectors that have adopted formal reliability approaches and standards.

The definition contains four distinct parts: a probability, which means there is some uncertainty; a function (or functions); a stated interval or time period; and stated or assumed operating and environmental conditions. This definition is very objective and specific, and in turn lends itself to the ability to assign quantitative numbers.

Terms that are frequently included with reliability, and often confused, are availability and dependability. The distinctions around reliability and availability are usually most frequent within the reliability and maintainability community and are not frequently misused (or even on the radar) among decision makers. From a technical perspective, reliability addresses the period leading up to the first failure in repairable systems. For repairable systems, availability addresses the period following the first failure. Reliability and Maintenance (R&M) professionals simply refer to availability in terms of "uptime" and "downtime." They focus on the aspects that improve the former and reduce the latter. In layman's terms, "inherent reliability" is often used to describe how well a system, asset, or part has been designed and built to meet its intended function.

Dependability is often underused in R&M circles, but is a powerful term that is usually synonymous with reliability when communicating with decision makers. Dependability is commonly defined as the probability that an item will meet its intended function during its mission. It is very similar to reliability in the sense that time and operating conditions are specifically limited to the mission. In terms of a NASCAR race car, over the course of an entire season the car may not be as reliable or as available as desired; however, what matters most is that it is extremely dependable over the four hours of its core mission that occurs on most Sunday afternoons.

For many general audiences, I draw the distinction of reliability, availability, and dependability to my wife choosing me as her husband. As a spousal partner, she determined that I was reliable – that is, I had a high probability of meeting her desired functions over her lifetime of living in the southeastern United States as a middle-class family. However, if you asked her how available I am she would probably say "somewhat limited, he seems to require a lot of downtime." Yet if you asked her if I were dependable, her answer would probably be "yes, he normally performs quite well when I need him to do so."

RISK

Risk has many definitions. For this reason, what is meant by the terms reliability and resiliency may be intuitively understood by decision makers. Almost everyone has their own working understanding of the term risk. It is almost certain that there is not a common understanding by a collective group of decision maker(s) and their support network of what is meant by the term risk.

One source of confusion is rooted in risk professionals themselves. Most decision makers would say that risk involves the potential for loss. However, the source of many heated technical debates among risk professionals is whether risk involves the potential for loss (a potential for gain would be an opportunity) or whether risk can be either a negative or a positive occurrence. Most decision makers are surprised that the dominant technical definition among risk professionals is that risk can be either positive or negative.

There is a basis for this technical definition dating back to Frank Knight's classic book, *Risk, Uncertainty, and Profit*, related to financial management (Knight, 1921). The treatment of risk as both a positive and negative has been roundly debated as being confused with uncertainty or variability. A close inspection of Knight's original treatise leads a reader to the opinion that Knight equated loss with the negative effects of both risk and uncertainty. The greater issue was distinguishing "risk" from "uncertainty." From a practical perspective, the decision maker typically associates risks with negative events, and normally risks are thought of in terms of a potential loss of value.

Another source of confusion related to a foundational understanding of risk is the traditional expression of risk among engineers and many other technical professionals as the multiplicative product of the consequences of failure and the likelihood of failure. The truth is that the product of the consequences and likelihood of failure is simply *one way* to express risk and is not the definition *of* risk. Decision makers and the public are often confused by engineering professionals who express risk in terms of some ordinal number on a scale of 1 to 10 or as some relative percentage.

Risk is defined by the international risk standard, ISO 31000, as "the effect of uncertainty on objectives." The Project Management Institute (PMI) has adopted the same definition as well as a number of other organizations. The Council of Supporting Organizations (COSO) of the Treadway Commission is one of the few standards that still maintains that risk is the possibility of a loss.

Risk is defined as the effect of uncertainty on objectives.

The notes provided in ISO 31000 related to the definition of risk are frequently poorly appreciated and often not well understood. The first is that an "effect" is a deviation from the expected—either positive or negative. One key reason that advocates of the "positive or negative" school of thought stand by this definition is that they believe both the potential positives and negatives should be tracked, say in a risk register, and should be treated as well as communicated.

To me, maybe the most important aspect of the definition and this note is "whose objectives?" and "whose expectations?". Objectives and expectations often depend on what one has to lose or gain, and therefore the implication is that risk is subjective in nature.

The second note to the definition further supports the subjective nature of risk by the recognition that "objectives can have different aspects and can apply at different levels." Therefore, risk not only is affected by what one may lose or gain but also by where one may sit in an organization. This has intuitive appeal because it often reflects what we observe in the real world. Most Chief Executive Officers have a different perspective than an employee on the front line of what is risky.

A third note is that risk is "often characterized by reference to potential events and consequences, or a combination of these," as previously mentioned. ISO 31000 describes 24 different ways that risk can be characterized or expressed. While this note to the formal definition in some ways may add confusion, the body of ISO 31000 also adds clarity that the characterization or expression of risk is not the same as the definition of risk.

A fourth note that adds meaning is "uncertainty is the state, even partial, of deficiency of information." If we go back once again to Frank Knight, his operative concepts were that risk is quantifiable and uncertainty is unknown or unquantifiable from a practical standpoint. Risk and uncertainty can be positive or negative. Again, Knight acknowledges that risk is normally thought of as a negative occurrence. Throughout *Risk, Uncertainty, and Profit,* he refers to risk in a negative context, although clearly this is not his stated intent.

Knight's ultimate point is that risk is quantifiable. If quantifiable, risk is then openly reflected in market prices in a free and efficient market. Risk can be reduced by acquiring more knowledge; however, uncertainty cannot be reduced with more knowledge. In modern discussions among risk professionals, what Knight would call uncertainty is often referred to as "inherent risks."

The fine points (and theories) related to risk and uncertainty are not mentioned or debated in modern risk standards such as ISO 31000. It is worth noting that Knight notes that few markets are indeed efficient, and therefore profit is the result of either *inferred or real* uncertainties. He reflects the real or inferred aspects when he states "we can also employ the terms "objective" and "subjective" probability to designate the risk and uncertainty, respectively, as these expressions are already in general use with a signification akin to that proposed." Additional discussions on probability, the nature of measuring it, and the degree to which it is subjective are found in later sections; however, the point to be reinforced here is that one's concept of risk is largely impacted by their subjective opinion of the uncertainty of future events.

There is one additional aspect related to risk that is tremendously important but that is not formally expressed in the definition itself. This important aspect—time— is often critically overlooked even by so-called risk professionals. Unlike reliability, whose definition clearly prescribes a time element over which a criterion can be measured (like hours of available operation), the time period associated with risk is left to the eye of the beholder. This further leads to the subjective nature of risk. Along with "whose objectives" and "whose expectations," the clarification of "over what period of time" is essential to the measurement, communication, and understanding of what we express as risk to decision makers.

RESILIENCY

Resiliency is the ability to return to the original form or state after being bent, compressed, or stretched. Resiliency has also been defined as the ability to recover or return to the desired state readily following the application from some form of stress such as illness, depression, or adversity.

Resiliency has been adopted by the environmental community in recent years as a substitute concept for the overworked term sustainability. Adaptive management, commonly defined as "learn as you go," has grown as a focus by the US Environmental Protection Agency (USEPA) to create more sustainable and more resilient environmental systems. In this context, resiliency has been further defined—and expanded—as the capacity for a system to survive, adapt, and flourish in the face of turbulent change and uncertainty (Fiksel, Goodman, & Hecht, 2014). This same alternative concept of improved resiliency through adaptive management is also an emerging concept in operational resiliency related to man-made systems such as machinery. In this case, resiliency is improved beyond its designed level through human learning and leadership. This shifting definition of resiliency is unfortunate because it further creates confusion among the public, decision makers, and technical professionals.

I like to describe the association of resiliency to reliability and risk in the following way. Reliability establishes the essence of performance of any system in terms of its likelihood of success, function, operating conditions, and time frame. Reliability should be in excess of 99%, and very near 100%, for most systems used by humans. The likelihood component of risk can be then considered synonymous with unreliability, and therefore, in turn, expressed mathematically as an additive function of reliability plus unreliability equals one. The tradeoffs then for risk and reliability become the costs, time, and quality associated with making something 100% reliable (which, in reality, there can never be 100% reliability in the face of uncertainty). However, the operative concept in the tradeoffs analysis is how much potential pain the decision maker is willing to accept for unreliability versus the costs associated with making something as reliable as possible.

Resiliency kicks in when unreliability happens – the risk has been realized. Said another way, resiliency is needed when a failure from the original system reliability has occurred. Resiliency is the ability to return to the original state— or the system design given desired functions, operating condition, and time period. Expectations are also important related to the time to return to normal.

Resiliency is most relevant for perceived rare or catastrophic events, namely those events outside what may be considered normal or predictable conditions. After all, normal or predictable conditions should be considered in the reliability boundary conditions of a system. The concept of rare events is therefore very important in making systems resilient and important in communications with decision makers.

Clear and concise definitions of reliability, risk, and resiliency are essential for effective communications and decision making.

A SYSTEMS EXAMPLE

I like to use what I refer to as "static" systems to explain fundamental aspects of reliability, risk, and resiliency. Static systems are comprised of primarily assets that do not move in the normal operating state and do not have significant chemical or biological processes. Infrastructure systems like roads, bridges, pipelines, and buildings tend to be static in nature. They make good introductory examples for discussing reliability, risk, and resiliency.

I was involved early in my career in the analysis of a collapsed bridge. The abbreviated background is that the bridge served as an overpass across a railroad track that was located below it. Following several days of heavy rain, the track bed and rails became unstable and the train derailed as it passed beneath the bridge. Several of the cars contacted with the foundation piles that supported the bridge deck. The piles that supported the middle spans of the bridge were made of timber. They snapped like toothpicks. The middle bridge spans collapsed onto the tracks and onto the derailed train. One person was killed and another person was severely injured when their cars plummeted into what had become a deep pit filled with twisted metal, concrete, and timber.

The bridge had been constructed during two time periods. The middle spans, which were the original structure, were approximately 30 years old and were the ones supported by unprotected timber piles. This original structure was from a period when the area was predominately rural. The outer spans, essentially two new structures on each side of the original bridge, were constructed approximately 20 years after the original bridge was constructed. The newer structures were constructed with much stronger pre-cast, pre-stressed concrete piles. The outer spans were used as emergency lanes and did not serve daily traffic.

As was customary with bridge expansions, the newer structures and the older (original) bridge were not tied together in any way. They were designed as independent structures. And that is exactly how their owner, the state department of transportation, and their design engineer, wanted them to be designed. Tying the two together would have meant any latent defects or vertical settling with the original structure would cause problems with the newer structure. Keeping them independent minimized the risk and potential liabilities of the old from affecting the new. In fact, the addition and the existing structure were analyzed so independently that the new did not even consider concrete barriers to protect the exposed timbers piles of the old, even though the new bridge used concrete rather than timber piles to support the spans.

From a reliability perspective, both the newer addition and the old structure could be considered theoretically reliable. At the time of their designs, both had high probabilities that they would safely carry vehicles over the railroad tracks (their primary function), under stated conditions (including normal weather conditions), for a stated period of time (usually 30 to 50 years for a bridge). However, it can be argued that the number of vehicles and number of trains assumed in the reliability statement for the old bridge was much different than when it was designed; it's operating conditions had changed significantly as it approached the end of its useful life. The new addition had not collapsed, even when impacted by the train cars, so it could easily be argued that it was reliable.

The problem is that the bridge system was not reliable. It experienced catastrophic failure. There was severe injury. There was death. It did not do what its current users wanted it to do, which was to allow them to travel safely over the railroad tracks. The missing piece was considering the bridge as a single transportation system at the time the new structure was added. The reliability statement would have looked much different in this scenario. And it is not likely in this actual scenario that one structure would have collapsed while the other was left standing. One descriptor for the design was that the new structure had been designed for integrity but the transportation system had not been designed for reliability.

The absence of a true reliability analysis of the bridge system as a unit is one source of blame. Another is the evaluation of risk. As stated earlier, if the definition of risk is the deviation from objectives, then the question should be related to whose objectives are considered. The highway administrator at the time of the expansion probably had few objectives related to correcting problems of the past or building more expensive bridges. The design engineer was likely following the design standards of the day that were intended to protect the public, and there was little incentive for him or his insurance carrier to take additional liability by fixing potential problems of the past. Some risk calculations could show that there was a very low probability in the past of a train derailing and knocking over a bridge. A risk-based analysis could show that the tradeoffs of spending extra money to prepare for a rare event, that would not likely happen during the useful life of the bridge, was not worth it.

The only thing missing from that evaluation of risk is what matters the most. This is the desire of the user. In reliability and quality circles this is referred to as "the voice of the customer." Most users of the bridge would say that there is no expectation to die from a bridge knocked over by a train when the primary function of the bridge to safely pass over the railroad track.

Resiliency provides a third perspective. Resiliency is the ability to return to normal after being stressed. Because structures are rigid, we normally think of their resilience in terms of being able to absorb shock or stress without collapsing. Clearly the old bridge was not resilient under its current operating conditions.

A reliability approach would have helped decision makers prevent a failure by avoiding the distress while a resiliency approach would have helped decision makers prevent a failure by thinking about how the structure would respond to the distress. In this case, either approach would have yielded a better transportation system than the one provided.

SUMMARY THOUGHTS ON RELIABILITY, RISK, AND RESILIENCY

- Definitions are critical to developing common understanding, building consensus, and making quality decisions.

- Most decision makers do not understand the basic terms —reliability, risk, and resiliency. When asked, most reliability, risk, and resiliency professionals cannot quickly cite a standard definition that can be easily understood. It should not be surprising that communication in this subject area is less than good.

- The most important aspect of this chapter is to have a quick, standard definition always on the tip of your tongue, cite it in your communications, and to some level be able to discuss the fine points of the definition if asked.

When engaging in verbal presentation, always state something to the effect of "by risk I mean…." The same goes for reliability and resiliency.

CHAPTER 2

OTHER CONCEPTS THAT CONFUSE

Another exercise that I do in workshops is to ask another simple question. By either an Audience Response System (ARS) or simple show of hands, I ask the following question, "What does the chance of rain mean?"

Four potential answers are provided:

A. The percent of area over which precipitation will fall.
B. The percent of time precipitation will be observed on the forecast day.
C. The amount of days similar to this one when it actually rains.
D. The amount of rain that will fall.

Over the years, I have found the responses in each session tend to be divided almost evenly between the first three answers. The only venue where I almost always get the one correct answer, and unanimously, is Palisade user conferences. Palisade is the producer of the probabilistic simulation tools @RISK and PrecisionTree. One should expect the technical users of these tools to get it correct. A few people in most sessions will chose the last answer, which of course is a throwaway, but one which I actually experienced many times while coaching baseball and debating the potential for rain-outs.

The correct answer is C. The point of the exercise is that most people do not understand probabilities.

This exercise is one I borrowed from well-known risk expert Gerd Gigerenzer, who also gives credit to Joslyn Nadav-Greenberg, and Nichols' "Probability of Precipitation" as his source. In his books and articles on risk, Gigerenzer demonstrates that the misconception of probabilities has led to a wide range of negative issues for the general population. Some of his most poignant examples relate to the diagnosis of breast cancer and AIDS. Many of the most educated professionals in society, medical doctors and attorneys, are used by Gigerenzer to demonstrate just how poorly the basic concepts of probability are misunderstood and miscommunicated.

PROBABILITY

There are three primary schools of thought related to why people almost universally misunderstand probabilities. These three schools of thought can be summarized: humans do not have a good intuitive understanding of numbers (innumeracy); humans do not naturally understand probabilities but can be trained to do so; and the issue is related most to our inability to communicate probabilities effectively rather than a natural human condition.

Behavioral psychologists who originated in the post-World War II era are the primary advocates of the first school of thought that humans are not naturally statistical thinkers. Daniel Kahneman and Amos Tversky were leaders in this field. The nature of their work was driven by the realization that subjective human judgement is biased, that we are too willing to believe based on inadequate information, that we are prone to collect and understand too little information, and that we are frequently misled by randomness. "People do not appear to follow the calculus of chance or the statistical theory of prediction" (Kahneman and Tversky, 1979). In more simple terms, as humans, we tend to reduce a vast majority of our decisions to simple rules of thumb because we are naturally too lazy and, to a comparable degree, unable to understand the numbers.

Gigerenzer straddles both the second and third schools of thought but comes down more squarely in the third school. To a point, he agrees that humans have a difficult time with statistical thinking. However, the problem is less with the human mind and more about how we teach, train, and communicate probabilities. Gigerenzer (2014) describes three skills that are required for becoming successful in risk-based communication and decision making: statistical thinking, rules of thumb, and the psychology of risk. Interestingly, he has proven that all are relatively better understood by fourth graders than by adults.

Gigerenzer (2014) concludes that to calculate risk is one thing, to communicate it is another. Risk is important to both decision makers and to experts, but the effective communication of risk is so rarely taught that misrepresenting numbers is the rule rather than the exception. His two solutions are simple: use frequencies (the number of observations out of the total class number) instead of single event probabilities (percentages), and use absolute risk (two in one thousand) rather than relative risks (reduction of six to four in one thousand). In other words, keep the communication simple by using whole numbers based on the observations rather than often misunderstood percentages.

The issues related to understanding probabilities has very practical ramifications. The example I use in the communication workshops come from asset management. When reviewing the work of others related to the subjective assessment of the likelihood of failure, I can "blow up" the previous assessment by simply asking the probability question in two different ways. When I conduct workshops and review results from facilitated sessions to gain an understanding of the likelihood of failure, I often use the following questions as examples: "What is the chance (or probability) that this piece of equipment will fail?" and "If you have 10 pieces of equipment just like this one, how many will fail?". The two questions have the same meaning. The first will almost always yield a higher answer than the second.

Most people do not understand what is meant by the chance of rain.
We have a poor instinctive understanding of probability.

This misunderstanding of basic probabilities creates problems related to expressing risk as a function of the likelihood of failure. Many experienced R&M professionals have experienced this misunderstanding. In recent years, research by the Department of Homeland Security and the Institute of Defense Analysis have documented this issue. Much of the misunderstanding can be understood through the basic concepts of measurement theory and scales, whose underlying foundations are the theories of representativeness, uniqueness, and the applicability of statistics to certain types of scales. The simplified conclusion related to the use of the likelihood of failure expressed as a probability is that risk professionals frequently get the underlying mathematics wrong. This further adds to the practical confusion in decision making related to percentages.

At a minimum, communicators of reliability, risk, and resiliency should report frequency (87 of 234 cases) rather than percentages (37.2%). Best practice is to present both whenever possible, and in most cases, it is relatively easy to show both.

Another added benefit of reporting frequencies is that it provides the decision maker with at least a practical insight into what could be at least loosely referenced as "confidence interval" or statistical significance—there is usually more confidence in 30% of 1,000 cases than 30% of 10 cases.

One final note that I am reminded to make before leaving the topic of probabilities. Reporting probabilities in terms of frequencies does not mean that a person champions frequentist statistics rather than Bayesian statistics. There are two major types of statistical inference: classical (or frequentist) and Bayesian. Frequentist statistics uses historical or demonstrated data to determine a probability distribution, point estimate, and interval estimate (confidence limits). Bayesian statistics quantitatively combines human belief (a subjectively-based probability distribution called the prior distribution) with operational or test data (an objectively-based probability distribution). Bayesian statistics uses new demonstration data in the form of a probability distribution to update a subjective probability distribution.

Both methods can generate either a point estimate or an interval estimate. Some organizations use both the frequentist statistics and the Bayesian statistics to convert historical performance data to a reliability estimate. Mentioned elsewhere in this book, Frank Knight was one of the first economists to use the term subjective probability when discussing risk and uncertainty. Jimmie Savage, perhaps the greatest statistician of the past two hundred years, is one of the greatest Bayesians. Both methods are viable.   Reporting probabilities in terms of frequencies is simply an effective way to communicate to decision makers. Reporting in frequencies does not mean that one is necessarily a frequentist or a Bayesian.

DECISION MAKING

Decision making is closely tied to the way humans think. Behavioral psychologists such as Daniel Kahneman cite what is referred to as System 1 and System 2 thinking. In communicating reliability, risk, and resiliency, it is very important to understand what type of thinking the decision maker is using in order for the message to be properly received.

System 1 thinking is categorized as intuitive, common sense, and *a priori*. System 1 thinking is represented by the hundreds of decisions we make each day to safely live our lives. In the domain of System 1, planning is limited, risk is evaluated by past experience and rules of thumb, and decision response is short and simple.

The example I use frequently is that of a tiger. If I walked out of a building and saw an actual tiger, my System 1 thinking would kick in and my decision would be simple – Run! But if I walked out of the same building and saw the Auburn University tiger mascot, Aubie, my immediate decision would be to say "Aubie!" and extend a high five. Neither response is well thought nor is it necessarily appropriate because the real tiger could be a toothless, tame one and Aubie could actually be a disguised terrorist intending to do me harm.

System 2 thinking is categorized as analytical, statistical, and *posterior*. System 2 thinking is characterized as the one or two life-altering or strategic decisions that some of us make once per week but most of us make maybe once per month. In the domain of System 2, planning is performed over the long term and involves both people and numbers. Risk is evaluated analytically and statistically. The decisions are thoughtful and measured.

The example I use frequently is a diagram of the six-step decision process championed by Stanford University and Strategic Decision Group. The six steps are frame the problem, develop alternatives, collect data, evaluate tradeoffs, establish logically correct reasoning, and commit to action. All steps are equally important. None should be skipped or evaluated to a lesser degree than the other steps. The final decision is only as good as the weakest step.

The single take-away from this discussion is that if your boss or boss' boss is using System 1 thinking to address what he or she believes is a relatively simple problem and you are responding in the world of System 2, then there will be a problem in the communication. The reverse applies as well. How we communicate should be driven to some degree by understanding the decision-making frame of the decision maker.

System 1 kicks in when we see a tiger. We run. System 2 is more deliberate and statistical. It is the realm of reliability, risk, and resiliency.

DEFINITION OF A DECISION

Another foundational component that needs clarity is the definition of a decision. The definition I prefer is again from Stanford University and the Strategic Decision Group. A decision is "an irrevocable (or irreversible) choice among alternative ways to allocate resources."

If you do not have an allocation of resources, then you have an "intended course of action." You do not have a "decision" if you do not have an allocation of resources. The simple example is going to a ball game or a musical at the end of the week. The intention or agreement to go to the event is the intended course of action; however, it only becomes a decision when the tickets are actually purchased.

Also, a decision is "irrevocable" in the sense that undoing the prior allocation may require allocation of additional resources. Resources may include things like time, money, and integrity.

From a communication standpoint, this is extremely important. I often find myself working with organizations who either have many meetings but find no resolutions or re-visit the same decision over and over. In the former case, none of the participants may have the authority to commit resources, and therefore a decision can never truly be made. In the latter case, the authority to commit resources may be present, but an allocation is not made. The result in both cases is the same—the long road of intended course of action. It is important that the resources that need to be allocated be properly conveyed if the communication is to be effective and if a decision is actually to be made.

WHY DO WE HAVE A HARD TIME MAKING STRATEGIC DECISIONS?

The answer to this question is a less obvious but critically important aspect of communicating reliability, risk, and resiliency to decision makers. Understanding the answer will make communications more effective. It is critically important because understanding why decision makers have a hard time making decisions will keep you from being shot as the messenger.

There are many academic schools of thought on this subject and as many as several dozen explanations cited from the researchers. As both a decision maker and as a consultant in a wide range of industries, I think it comes down to two things that continually pull in opposite directions. To depict the push and pull of any strategic decision, I usually use the example from the movie *Animal House* where a devil and an angel, both played by the late actor John Belushi, sit on opposite shoulders of a decision maker.

The first thing that makes it difficult to make a decision is that most strategic decision makers spend years climbing through the operational ranks. In the operational ranks, decisions are primary based in System 1 thinking; System 1 thinking consists of the tens or hundreds of transactional decisions we make every day. They usually require the allocation of small amounts of resources

When the decision maker moves to the top or above the operational ranks, then strategic decision making becomes dominant, and System 2 thinking is required. For strategic decision making, a single decision may take weeks or months to make and is dependent of various types of information to better evaluate complexity and uncertainty. The decision maker is subconsciously uncomfortable because of the time required to make decision. After all, for most of their careers at the operational level, tens or hundreds of decisions had to be made each day. The need to make a decision and proceed overpowers the need to make a good decision based on analytics and some form of a decision quality process. There is a personal relief to have done the job and put the decision behind us.

If this is the push, then the pull is the subsequent feeling of guilt if the action taken adversely impacts someone. The example I most often use is from Kahneman and Tversky. It basically describes two investors, Paul and George. Paul owns shares in Company A. During the past year, he considered switching to stock in Company B, but he decided against it. He now learns that he would have been better off by $1,200 if he had switched to stock in Company B. George owns shares in Company B. During the past year, he switched those shares to shares in Company A. He now learns that he would have been better off by $1,200 if he had kept his stock in Company B.

The question that Kahneman and Tversky posed in their research is "Who feels greater regret?" The responses were overwhelming with 92% of participants responding George (who took action) and only 8% responding Paul (who did nothing). Research of this type has been performed in numerous forms and by numerous people but the conclusions are all the same. As humans, we prefer inaction to action if the action ultimately creates pain to another human being or to ourselves. We somehow feel less responsibility and less guilt by our inaction.

So where does this leave communicators of reliability, risk, and resiliency? Simply put, our decision makers at some point are ready to pull the trigger, and at the same time, if their shot has unintended consequences, they will look to pass the blame. While our role as reliability, risk, and resiliency professionals is that of trusted advisor and messenger and not the ultimate decision maker, understanding the push and pull that drive our decision makers is important to getting them to a decision. Understanding the continuum is also critically important to surviving in our role.

Balancing the need for action and minimizing the blame for harming others makes strategic decision making difficult.

CASE EXAMPLE: POLICE SHOOTINGS

Gigerenzer used topics like breast cancer and AIDS as attention grabbing examples a decade ago. I use a modern example as an attention grabber: Do Police Kill More Blacks than Whites?

I cite numerous sources in discussing this example. Some sources use percentages, some use frequencies, and some use both. Of course, part of the issue is the population (all people, white people, black people, people in a geographic area) to which the sample of people shot by police are compared. The Washington Post does a solid job of presenting the data in many ways. Their presentation in turn helps the decision maker to understand better how the data has been converted to information. It also provides different formats that resonate with different types of decision makers which makes the communication effective.

The other aspects of the case example that I highlight are related to what decisions need to be made, are they tactical or strategic, and how does an individual reasonably evaluate their own personal risk based on the data. To a large part, the risk associated with the issue of whether police kill more blacks than whites depends on where you sit and is therefore subjective to a great degree. So too is the frame of either a tactical or strategic decision that needs to be made as well as the allocation of resources that is required with that decision.

All of this is intended to drive some good discussion around risk, percentages, and decision making. We are not trying to reach consensus on the topic. The overall take-away is to pay attention to how the data is communicated to support the narrative that is trying to be conveyed.

CASE EXAMPLE: PACIFIC WEATHER PATTERNS

One day while flipping through cable news channels, I ran across an interesting report. Scientists are now able to more accurately predict droughts in the Midwest United States two months in advance based on weather patterns in the Pacific Ocean. This obviously had huge positive implications from a number of perspectives. As a sailor and one who has studied and depended on weather patterns for a long time, it also sounded too good to be true.

I started listening carefully for the details. Finally, at the end of the report, it was mentioned somewhat casually that the prediction rate was better than 50 percent two months in advance and that more research money was needed to improve the forecast period beyond 60 days. Yes, down deep in the communication was the truth—researchers had developed a tool with the same accuracy as flipping a coin.

The communication aspect fascinated me in terms of where the misrepresentation had occurred. I pulled down the research and dove into its details. What I found was a whole lot of research and a really deep statistical analysis that was used to develop a prediction model. The researchers used the statistics in a way that confuses a general reader; however, to their credit, they referenced the accuracy in terms of its equivalence to a coin flip, albeit embedded in the text of the report. Buried in the statistics and the percentages was the fact that their forecast model was based on underlying data that did not include the great drought of the 1930s; nor was the model validated to predict the extreme event in that period. It was unclear who did the final journal article; whether journal editors embellished the findings to sell more journals; or whether the researchers embellished it a little to gain more future research funding. What I highly suspect is that the national cable news station just picked up the article from a national scientific journal, assumed it to be truthful, and ran with it.

The take-away in this example is to be wary of the statistics of research and the way the technical information is conveyed to the public and decision makers. For a decision maker concerned about the risks of a major drought in the Midwest, pumping more resources into the black box forecast tool would likely appear to be a good idea. Understanding the meaning of the probabilities and statistics probably leads to a different decision.

SUMMARY THOUGHTS ON OTHER CONCEPTS THAT CONFUSE

- Most decision makers do not understand the basic concepts of probability. There are three major schools of thought around why the general population does not understand probabilities, and each has a number of recommended approaches for communicating probabilities. The single best practice is to communicate frequency (87 of 234 cases) rather than percentages (37.2%). If communicating percentage, frequency should also be included. In most cases, it is relatively easy to show both.

- There are two types of thinking processes associated with decision making. System 1 is intuitive, common sense, and most appropriately applied to the tens or hundreds of tactical decisions that individual beings must make at home and work each day. System 2 is analytical, statistical, and most applicable to the types of work performed by reliability, risk, and resilience professionals. Both the analytical approaches and the communication approaches must be consistent with the type of problem that the decision maker believes is being solved. And to be effective, the communication must be consistent with the decision maker's frame rather than the analyst's frame.

- A decision is "an irrevocable choice among alternative ways to allocate resources." If you do not have an allocation of resources, then you have an "intended course of action." You do not have a decision if you do not have an allocation of resources. Most decision makers and technical professionals alike have a foundational definition of a decision. Communication cannot be fully effective if we do not know where we are going, or what the desired result should be.

The technical professional and the decision maker should be aligned with the type of decision that is being made. The type of decision drives the associated thinking processes (tactical or strategic) and the types of analysis that is required. Being aligned is a first step in framing the problem. On the back end this alignment will keep the communications relevant to the decision at hand and will be the first step in making sure that the decision maker does not short the messenger.

CHAPTER 3

GRAPHICAL EXCELLENCE

Communicating reliability, risk, and resiliency normally requires conveying a lot of data in a concise manner. Graphics are the most effective form of telling the story that large sets of data provide. This chapter is intended to provide a textual discussion of the use of graphics; however, it does not provide the many visuals that are included in the workshops or that can be found on-line from other sources. The concepts associated with graphical excellence are foundational to the effective communication of reliability, risk, and resiliency to decision makers. Therefore, this subject appears early in the workshops and in this book.

I normally begin this session of a workshop by making the point through Anscombe's Quartet, which was first formulated in 1973 by statistician Francis Anscombe. The quartet demonstrates four visually different data sets with the same mean, variance, correlation, and line of linear regression. The point is that visualization is more important than simply using the numeric expression of the statistics.

Renown graphical excellence expert Edward Tufte uses Anscombe's Quartet as lead-in to his highly-regarded book, *The Visual Display of Quantitative Information*, to emphasize the importance of graphics. Anscombe's Quartet demonstrates the importance of graphing data before analyzing it, the effect of outliers on statistical properties, and most important, the limitation of trying to communicate statistical data through words or symbols.

TUFTE'S EXPERTISE

Edward Tufte describes graphical excellence as:

- The well-designed presentation of interesting data—a matter of substance, of statistics, and of design.

- Consists of complex ideas communicated with clarity, precision, and efficiency.

- That which gives to the viewer the greatest number of ideas in the shortest time with the least ink in the smallest space.

- Nearly always multivariate.

- Requires telling the truth about the data.

At this point in a workshop I normally have a breakout session for the critique of visuals. Some of these I borrow from Tufte's books and his standard workshop on the subject, while others I have collected from other sources and my own real world experiences. I use some of the classics from Tufte's material: Minard's *Napoleon's March to Moscow*; Garofalo's *The Genealogy of Pop/Rock Music*; *Matterhorn Landeskarte der Schweiz*; Gallagher's *Spotting A Hidden Handgun*; and Galileo's *Theorica speculi concavi spaerici*.

Tufte's Six Principles of Graphical Integrity are the best single summary for graphic excellence that I have found:

1. The representation of numbers, as physically measured on the surface of the graphic itself, should be directly proportional to the numerical quantities represented.

2. Clear, detailed, and thorough labeling should be used to defeat graphical distortion and ambiguity. Write out explanations of the data on the graphic itself. Label important events in the data.

3. Show data variation, not design variation.

4. In time-series displays of money, deflated and standardized units of monetary measurements are nearly always better than nominal units.

5. The number of information-carrying (variable) dimensions depicted should not exceed the number of dimensions in the data.

6. Graphics must not quote data out of context.

These six points are so relevant and powerful that I provide multiple visual examples of each in the workshop setting. Obviously, Tufte's works and website provide ample examples. Older books such as Huff's *How to Lie with Statistics* and

Monmonier's *How to Lie with Maps,* as well as numerous more modern references, also provide additional examples. The reality is that examples are all around us in our own lives. I usually prefer to use these as real-life examples when possible.

OUR DUTY AS RELIABILITY AND RISK PROFESSIONALS

Graphical excellence has been described earlier in this chapter. Conversely by way of example, graphical manipulation has also been described. It is relatively common to encounter someone in a workshop that finds the entire line of discussion to be unethical or even offensive. It is not intended to be. Understanding the potential for manipulation is important in making sure you communicate properly and effectively.

At this point, it is important to take count of our duties in communicating information to decision makers. As reliability and risk professionals, our responsibility should be to express a given situation as fairly and objectively as possible. If for no other reason, this is important because the communication you provide will be passed "up the line" and each audience is likely to be a little different. Risk and reliability professionals are not communications experts, marketing professionals, or political operatives. Leave the manipulation, if needed, to those specialists. Above all else, for reliability and risk professionals, ethics, objectivity, honesty and duty do matter. Truth provides enough complexities and peril without throwing in an additional aspect of manipulation.

CORE GRAPHICS, THEIR HISTORY, AND MODERN USE

Leaders in the field of the graphical presentation of data include Tufte, J.H. Lambert (1728 – 1777), William Playfair (1759 - 1823), and John Tukey (1915 – 2000). One important point is that the history of the use of graphics is relatively new in comparison with the history of civilization. Another important point is that every type of graphic was driven by the necessity to improve communication and understanding. A third point is that the original versions also lead us back to the essentials of what is important with each, and in doing so, provides clarity.

Data maps originally developed in the form of geography-related maps and plots. As early as 1100, the Chinese had extremely good map making, also called cartography. It took cartography several more centuries to develop in the western world. Western maritime charts, including Edmond Halley's 1686 chart, and Dr. John Snow's 1854 map used to solve the cholera epidemic in London are two of the first historical and most powerful examples of cartography. Snow's example has also been used as an early example of the use of maps to show data clustering, although it would take nearly 100 more years for such data maps to be used statistically instead of simply as a method to show clustering.

Line and time series plots were created in the late 1700s. Joseph Priestly's *A Chart of Biography* (1765) is perhaps the best early example and really was the outcome of the history of religion and philosophy. J.H. Lambert is credited with the prolific use of line and time series plots in scientific journals beginning around 1779. William Playfair used this type of graphic to communicate economics, politics, and science in his *Commercial and Political Atlas* (1786).

Bar charts and histograms were also first used by Playfair in *Commercial & Political Atlas* (1786). According to Tufte, 43 of 44 graphs in the Atlas were time series and one was a bar chart. Playfair apparently had reservations about bar charts since they did not state the time relationship over which the data was collected.

Pie charts were first used by Playfair in *The Statistical Breviary* (1801). Like bar charts and histograms, they lack a time element. Pie charts in some ways are pleasing to the eye but are also technically limited in types and depth of data they can convey.

Box plots and box and whisker diagrams were not developed until the mid- to late 1900's. Tufte attributes "range bar" to Mary Eleanor Spear in *Charting Statistics* (1952). Tufte attributes "box plot" to John Tukey, *Exploratory Data Analysis* (1977), and the modern box and whisker diagram is attributed to Tukey.

The box plot, or box and whisker diagram, is analogous to a cat's face. Most of the action occurs within the face of the cat; however, the whiskers define the outermost extent of the cat's body.

Data plots and maps, line and time series plots, bar charts and histograms, pie charts, and box and whisker diagrams are core graphics. They are used universally in a wide range of fields of studies. All can be produced by hand. All can be produced in common software like Microsoft Excel or other commercially available database software.

COMMON RELIABILITY GRAPHICS

Most reliability professionals are familiar with Total Quality Management (TQM). This methodology originated in the 1950s and describes an organizational culture that provide customers with the products and services that satisfy their needs. The culture requires quality in every aspect of operations with every employee sharing equally in assuring that processes are done correctly the first time.

Perhaps the most important to the communication of reliability are the Seven Tools of Quality (American Society of Quality, 2017). Most have their foundations in the previously described core graphics.

A check sheet can be described as a structured form for collecting and analyzing data. It has been used throughout time in a wide variety of fields and for a wide variety of purposes. It is used when data can be observed and, in many cases, is needed when observations are needed repeatedly. Additionally, it is a common quality tool in the production process for collecting information related to defects. Last, it is commonly used in the construction industry as a "punch list."

Graphic Type	Description	Creator (year)
Check Sheet	list of needs or actions	Unknown
Control Chart (Time Series)	if process is in state of statistical control	Priestly/Lambert/Playfair (late 1700s)
Scatter Diagram	x-y graph of two variables in a data set	Lambert (1765)
Histogram	frequency diagram	Playfair (1786)
Pareto Chart	bar chart with cumulative line graph	Pareto (1906)
Flowchart	representation of process or algorithm	Gilbreth (1921)
Fishbone Diagram	cause and effect	Ishikawa (1940s)

The medical industry also uses a checklist in a number of ways, ranging from giving basic physical exams to using it as a quality assurance tool in surgical procedures.

A control chart is an example of a time series plot that is used to determine if a process is in a state of statistical control. Control charts are important beyond their classic use in evaluating the production process. Reliability requires that an item perform its intended function over a period of time.

Using a control chart in combination with statistical process control is a fundamental method to evaluate whether the intended function of any system is being met over time.

A scatter diagram is effective in analyzing and communicating the relationship between two variables. It is highly powerful in visualizing statistical correlation. Anscombe's Quartet, discussed early in this chapter, is an example of the effective use of a scatter diagram for communication.

Histograms are used to show frequency distributions related to how often different values in a data set fall into a certain range or "bin". Technically histograms are used when data is described as continuous and bar charts are used when the data is classified as integer. In general, histograms are the most difficult of the Seven Tools of Quality for decision makers to understand. Therefore, their use to produce effective communication is often limited.

The Pareto diagram is a form of bar chart. Created by economist Alfredo Pareto, the diagram is most commonly used to reference the "80-20" rule. However, its intended use is not specific to only "80-20" but rather the depiction that a large number of outputs can be usually attributed to a small number of inputs. It is considered by many to be the fundamental tool of reliability. Techniques such as Defect Elimination are based upon it.

The Pareto diagram, or the depiction of the "80-20" rule, is the "first graphic of reliability." It is a special form of the bar graph.

The flowchart is a modern reliability tool that separates functions or sources of data into groups. The relationships between the functions can then be connected to show inter-relationships. Flow diagrams are common in process engineering as well as workflow analysis and business process mapping. Organization charts can also be considered a form of a flowchart. If kept relatively simple and precise, flowcharts can be highly effective communication tools when working with decision makers; however, they are poor communication tools if overly complex in the form of process engineering diagrams or process mapping in the form of "spaghetti" diagrams.

The fishbone diagram is also known as a cause-and-effect diagram or the Ishikawa diagram, after its founder. It identifies the many possible causes (the "bones") that produce an effect or problem (the "head"). It is a powerful tool to use in structured brainstorming sessions or to stimulate fresh thinking on a problem.   It is a core tool in root cause analysis.   As a communication tool, keep the focus on the higher-level aspects of the bones and the head.

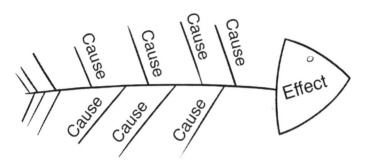

The fishbone diagram is a powerful tool for cause-and-effect analysis.

RISK PROFESSIONALS

Risk professionals use the same graphic techniques that have been described previously under the topics of core graphics and common reliability graphics. These graphic techniques include scatter diagrams to show correlation between variables, histograms and bar charts to show frequency relationships, and time series plots to show consequences (especially number of occurrence or total dollars) versus time. The heat map is relatively unique to the risk profession and therefore worth further discussion.

Heat maps are a type of scatter diagram. Typically, the consequence of a negative event (failure), and the likelihood of a negative event (failure) is plotted on a two-axis plot. At least three zones are usually depicted. The zone of high consequence of failure and high likelihood of failure is usually depicted in red or another hot color, low consequence of failure and low likelihood of failure as green or another cool color, and the intermediate zone as yellow or another intermediate color. The hot and cold depictions of risk are the obvious source of the term heat maps.

Heat maps are often provided with either integer numbers or no numbers at all. In many cases, they can be misleading because their graphical depiction is based on interpreted data (like scoring based on subjective perceptions with Likert scales) rather than directly measured data. Similar to pie charts, they should be regarded as pleasing to the eye but technically limited in types and depth of data they can convey. Nevertheless, they are commonly used and can be used as an effective communication tool to decision makers when used honestly and to make simple points.

ART AND CHARTJUNK

A common question in many workshops is related to the amount or degree of the use of colors, pictures, figures, animations, videos and related forms of media. My fundamental answer is "very little." As reliability and risk professionals, we are making presentations to convey data and information as fairly, objectively, and concisely as possible. We are not the ultimate decision maker. Our goal is not to manipulate, sell a product, or win a vote.

The more complicated answer is that we also must be effective in our communication. We should be less controlled by rules and most controlled by doing what it takes to get the decision maker to receive our message effectively. In practice, this means that most of the time we stick to the fundamentals. At other times, we bend or break the rules.

The fundamental approach is to keep it simple, clean, and concise. Eliminate as much of the noise and clutter as possible. Edward Tufte's third principal of graphical integrity is to show data variation, not design variation. That principal sums it up well.

One of our primary hurdles from a practical perspective is that most software, including simple tools like Microsoft Excel and Power Point, have much more power and variety than we actually need as professionals trying to communicate reliability, risk, and resiliency to decision makers. Combine this with a corporate graphics or publishing department, and now we have tremendous pressure to be creative and use all of the resources we have at our disposal. Soon we find ourselves creating "art" rather than communicating data and information.

The term "chartjunk" refers to the unnecessary distractions in the graphics, particularly those often produced in Microsoft Excel, that create noise. I am unsure if Tufte coined the phrase, but certainly he has championed the fight against chartjunk. For me, chartjunk is just a sub-example of falling into the trap of creating art rather than focusing on clear communication of the data.

A bit of fair warning on interaction with the corporate graphics and publishing professionals. They do not like the term chartjunk because it implies they are creating junk. And they do not like the reference to art because they know they are not being paid to create art (even though many were trained as artists or consider themselves artists outside the corporate setting). Most are trained in the art of persuasion and marketing manipulation. They are judged by the peers and superiors, who are also trained in the same ways. They ultimately report to the same decision makers that you do but are in different organizational silos.

Understanding is the common bridge between communications and learning. Ruth Colvin Clark is a renowned expert in workforce training and development. She believes that graphics are key tools for minimizing information overload as long as they themselves do not become the source of overload. Clark distinguishes between decorative visuals (used to break up heavy details and as "feel good" devices) and explanatory visuals (used for explaining the data). The decorative visuals can be effective but also tend to create additional load on the human brain. Explanatory visuals should be our focus as technical visuals. Keep it simple.

Clark also points out that visuals are not always needed. The myth that some people are predominately visual thinkers and must learn through images, graphs, and charts is one that she debunks from educational research. Another point she makes is that visuals are not needed to the same degree in workshops with other technical professionals as they are with non-technical professionals. In my words, technical professionals can often see a process and instrumentation diagram (P&ID) in three dimensions while a non-technical will have no understanding of depth and distance without some form of geospatial diagram (and preferably in three dimensions).

I give three pieces of advice to practitioners: personally control your supporting graphics as much as possible; educate and train yourself on the effective use of graphics; and, when working with corporate graphics and publishing specialists, do so in a collaborative manner and from the perspective of expressing the messages you are effectively trying to deliver. Your corporate graphics experts are always going to assume the high ground in any related debate. Fighting them is a war you cannot win. Fighting them is also a major distraction from effectively communicating your data and information to decision makers.

COLORS

The use of colors is a common question as it relates to the subject of graphical excellence. My normal response is that this is both an art and a science. There are many books, workshops, and courses on this subject (normally taught by marketing professionals). I recommend a deeper dive into this area by any technical professional; more discussion is also provided in Chapter 6 related to noise since the potential impact of colors extends to other communication tools beyond graphics. I provide a couple key practical points and observations below.

First, use colors primarily for emphasis. Stay with greyscales and pastels to the extent possible and use bolder colors to draw the audience to the key point you are trying to make. Remember that people who are colorblind will have problems interpreting graphics that are highly dependent on color. Colors often appear differently after reproduction or when projected onto a screen which can result in negative impressions. The first goal is to keep the graphics neat, clean, and simple. The second goal is to use colors appropriately to emphasize the key message.

Second, black, dark blue, and white are always effective. Black is psychologically associated with finality and when used with greyscale can be effective in adding emphasis. Dark blue is associated with calmness, credibility, and a business-like approach. White is a great background as well as great as text on black or blue backgrounds.

Red, green, yellow, orange and purple are suitable for adding emphasis. As a rule, they should not be used as primary colors when communicating reliability, risk, and resiliency to decision makers. As primary colors, they have high potential for creating distractions related to their use and in turn take the audience away from the underlying message. Also, they have high potential to create noise.

The following summary comments are provided in terms of adding emphasis. Red stimulates a strong emotional response and is normally associated with loss. Red can create excitement and attention to detail. Yellow is associated with optimism, is normally attention-grabbing, and stimulates mental activity (one reason it is a good color for high-lighting). Orange is a blend of the positive aspects of red and yellow, and therefore associated with newness, freshness, and youth. Green is associated with nature, vitality, and wealth. Green is an effective color for helping generate warmth, collaboration, and feedback. Purple is associated with nobility, royalty, and mystery.

A single negativity for each of red, yellow, orange, green, and purple is their potential to be reproduced or broadcast poorly and become distractions. From a negative psychological perspective, red can be associated with danger and stimulate a negative interpretation. Too much yellow creates a literal and metal glare. Too much lighter green is similar to yellow (dark green is usually safe). Orange is not a traditional business color and may give the impression of being too new or unproven. Purple is traditionally associated with whimsy and never mentally provides the professionalism or confidence that a black or dark blue produces.

Colors do matter, and that is one reason the question is often asked. The bottom line is to default to black and dark blue as the base color in combination with white. This is consistent with the serious nature of communicating reliability, risk, and resiliency to decision makers in a fair and objective manner. More importantly, using black or dark blue as the base color provides the greatest potential for the audience to focus on the data and information rather than on variation in the graphical design.

SUMMARY THOUGHTS OF GRAPHICAL EXCELLENCE

- Graphics are intended to summarize large quantities of data and should assist in providing unique insights. Statistics, expression of data using symbols or equations, and tabular data are not enough to communicate effectively to decision makers.

- Graphics are not always required to explain the data. Use data tables where the data is not complex or in large volumes. Graphics can also be a form of distraction when not done effectively. Most technical professionals are not well taught or well trained in graphics. Default computer programs and the corporate marketing department usually do more to reduce graphical clarity than improve it. Avoid "chartjunk", too many colors, and making a graphic a piece of "art."

- I discussed a range of graphics, their histories, and their use. Data plots and maps, line and time series plots, bar charts and histograms, pie charts, and box and whisker diagrams are core graphics that can be used for communicating reliability, risk, and resiliency to decision makers. All can be produced by hand using graph paper or on white boards. All can be produced in common software like Microsoft Excel or other commercially available database software.

- Edward Tufte is the single best reference for graphical excellence. For Tufte, a well-designed presentation of data is a matter of substance, statistics, and design. Graphical excellence requires telling the truth about the data. His work, including his Six Principles of Graphical Integrity, is referenced for further study.

CHAPTER 4

THE ROLE OF ETHICS

Telling the truth about the data is the most important aspect of communicating reliability, risk, and resiliency to decision makers. A fundamental question is related to the concept of truth and associated concepts such as objective, fair, and honest. Similar to basic definitions and graphical excellence, understanding the role of ethics is foundational to the effective communication of reliability, risk, and resiliency to decision makers. Making allowances for the possibility that other people may see the concepts differently than the communicator is a meaningful consideration.

In the workshop setting, I tell participants that this module is not intended to explain the deeper meaning or values of life. Nor is it intended to be a university course or full curriculum on the topic of ethics. In shorter, 30-minute to 1-hour presentations on communicating reliability, risk, and resiliency to decision makers, I half-jokingly tell the audience I am going to boil everything they have ever learned about ethics into one slide.

The truth is that I can indeed boil it down to one slide. And I do—every time I discuss communicating reliability, risk, and resiliency. It is that important.

To break the seriousness of the topic I normally provide the comparison of my work in Texas and Illinois. I tell the audience if I go to Texas and start talking about the impacts of climate change on decisions related to infrastructure and the environment, I will get fired. If I go to Illinois and don't talk about it, I will also get fired. Everyone laughs because the truth hurts.

So how do we decide what to leave in and what to leave out when it comes to communicating to decision makers? After all, the stakes are high for them. They are often high for us too as the communicator (sometime the messenger gets shot). The answer is that our ethics should make the final determination.

THREE TYPES OF ETHICS

Ethics can significantly impact the direction and shape of technical presentations. A traditional approach groups ethics into three main categories: virtue ethics, consequential ethics, and deontological ethics. Obviously, anyone who has taken a course on ethics realizes that it is a lot more complex. However, in simplicity there is usually clarity and in turn usefulness for practitioners.

COMPLEX   SIMPLE

In simplicity there is clarity and usefulness.

Virtue ethics describes the character of a moral agent (right and wrong, good and bad) as a driving force, and is used to describe the ethics of Socrates, Aristotle, and other early Greek philosophers. Aristotle described virtue as courage, temperance, liberality, magnificence, magnanimity, proper, ambition, patience, truthfulness, wittiness, friendliness, modesty, and righteous indignation. On the other extreme he framed vice as rashness, licentiousness, prodigality, vulgarity, vanity, ambition, irascibility, boastfulness, buffoonery, flattery, shyness, and envy.

In a similar frame of thought, the Apostle Paul in his letter to the Galatians described good as love, joy, peace, patience, kindness, faithfulness, gentleness, and self-control. On the other extreme he characterizes bad as fornication, impurity, licentiousness, idolatry, sorcery, enmity, strife, jealousy, anger, selfishness, dissension, party spirit, envy, drunkenness, carousing, and the like.

The point is that virtue ethics are based on the foundation that there is an absolute right and an absolute wrong. However correct or however noble, the reality is that it is very difficult to bring diverse decision makers to a singular allocation of resources if the underlying argument is right versus wrong, good versus bad, or virtuous versus non-virtuous. Virtue ethics are part of any decision, but finding virtue often takes a lifetime or more and many real-world decisions cannot wait on the journey.

Consequence-based ethics, or Consequentialism, refers to moral theories that hold that the consequences of an action form the basis for any valid moral judgment. From a consequentialist standpoint, a morally right action is one that produces a good outcome or consequence—in other words "the ends justify the means."

Consequence-based ethics received their primary development in the latter 1800's. John Stuart Mill and Jeremy Bentham were given substantial credit for this development. Interestingly, Mill and Stuart are also credited in economics with the creation of Marginal Utility Theory (MUT). In MUT, it is held that it is not the absolute value as described in Utility Theory that motivates the decision maker, but rather the incremental or marginal utility. In MUT, it is the comparative differences that matter most as well as the incremental pain.

The example I use to describe consequence-based ethics is the Affordable Care Act, also known as Obamacare. This usually both wakes up the audience and stirs some passion. My point is not about the ultimate value of universal health care but rather about how the issue was decided. One side believed that a 2,000-plus page piece of legislation should be read and understood before it was passed. The other side believed that taking the time to read and understand the bill was not important since those consequences were ultimately small compared to the overall value of universal health care. In other words, the ends justified the means.

Deontological ethics is an approach to ethics that determines "goodness" or "rightness" by examining acts, or the rules and duties that the person doing the act strove to fulfill. Immanuel Kant's theory of ethics is considered to exemplify deontological ethics. Deontological ethics holds that the consequence of actions do not make them right or wrong, rather it is the motives of the person who carries out the action that makes the actions right or wrong. Deontological ethics is in direct contrast to consequential ethics, and places priority on full disclosure and "treating others in the manner in which you would wish to be treated."

Reliability, Risk, and Resiliency communications should be duty based.

The legal requirements and code of ethics for many professions, namely engineers and physicians, are aligned to deontological ethics theory. Of the seven fundamental canons of ethics for engineers, three are of particular interest to communicating reliability, risk, and resiliency to decision makers. The first and foremost canon is engineers must hold paramount the safety, health, and welfare of the public. A second canon is engineers shall strive to comply with the principles of sustainable development in the performance of their professional duties. A third canon is engineers shall issue public statements only in an objective and truthful manner (American Society of Civil Engineers, 2016).

The term "hold paramount" in the first canon means "definitively above all others." It is a high standard among the universe of professions and their related ethical standards. Such a canon is important for engineers because it means the order of priority for the public communication of technical information is first and foremost the public good; next, the interest of their clients and employers; and finally their personal good.

This triangle of ethics is merely a framework and not an absolute. As previously mentioned, a number of ethical theories have been grouped under each of these canons. The reality is that many are not as "pure" as described in the simplified framework. Most individuals and groups do adhere predominately to one form of the three and secondarily to one of the others.

One example is professional engineers. Engineering codes of ethics are predominantly deontological, stressing and enforcing engineering duties. However, engineers often embrace consequential ethics in the form of solving problems that focuses on a results-orientation such as benefit-cost analysis. The primary ethical focus is duty-based and yet at some point in the trade-offs analysis, the final decision is made with a consequence-based focus. These potentially conflicting perspectives may often appear to be in conflict. Technical communication must therefore be precise, accurate, and representative. This underscores the value of a written form of ethics, which for some technical professions is enforceable by law.

THE ABSENCE OF AN ENFORCEABLE ETHICAL CODE

Laws and regulations at the local, state, or federal level often influence the ethical practice of professions through licensing laws. Obviously in these cases, the courts provide a system of enforcement and, in turn, some determination of fairness with respect to the communication of technical information. However, in many cases, the laws and regulations are neutral. This leaves the enforcement of an ethical standard to professional societies or associations. The consequences of an ethical violation in these cases is largely directly proportional to the status and visibility of the professional society or association in the eyes of its members.

I want to give a word of fair warning at this point. The public generally has more confidence in a licensed professional, whose code of ethical conduct is reinforced by law. In many cases, the public is not aware of whether an unlicensed professional—for example a certificate holder—has a legally enforceable ethical standard or whether it is merely enforceable by a professional society or association. For practical purposes, the ethical standards are only as good as the actual enforcement that is associated with them. The enforcement of ethics varies widely in most states and professions, even where the profession has legal licensing requirements.

But what about communication of technical information from people who are not legally bound by a code of ethics, including the general categories of researchers, academics, and scientists? A number of experts and organizations have called for a universal code of ethics from different fields, most recently biomedical engineering and related research (Vallero and Vesilind, 2007). In addition to the politicization of science, it is also a common dilemma that the ethics of business are quite different and less stringent than many of the underlying professions that encompass them.

One example related to the communication of risk and reliability that I often cite is related to my role as an appointed member of North Carolina's Environmental Management Commission (EMC). The EMC is the environmental rule making body for the state, and as such, the commission is often provided with technical reports that inform our rule making. One such report was related to the potential in-situ treatment alternatives for nutrients that had contaminated certain lakes in the state.

An initial report was prepared by the staff of the North Carolina Department of Environmental Quality and included advance material to commissioners and the public on the EMC website. The report did a good job of canvassing the possible range of in-situ treatment technologies and concluded that none were viable for the issue at hand in North Carolina. It also made the conclusion that protection at the source was the best method of resolution. The report was pulled from the agenda by DEQ on the day before the meeting.

One month later, a new report appeared. This one softened some of the technical discussion of certain treatment technologies, concluded that there were some in-situ technologies that had potential, and made recommendations for the funding of pilot programs for at least one of those options. No new data had been collected and no new information had been introduced; the report had simply been re-written to communicate the technical information in a way to satisfy certain political interests.

Obviously, this raises questions on the ethics related to this communication. The practical answer was that there were none. A licensed professional did not sign or seal either report. For that matter, none of the staff who were associated with the report were licensed in any way. It was true that all held at least four-year degrees, but all of these degrees were in either general science or business. Mid-level staff had done a good job of preparing the first report, and somewhere in the organization tree upper management had decided to present it differently. The official response was that DEQ had simply reached a different opinion after review of the first report.

The EMC saw it much differently. The first report was credible—there was no new information to justify the new findings in the second report—and quite frankly, the second report had a more biased tone than the first report when the two were compared side by side. Perhaps most importantly, the first report had been introduced in the public domain, which meant under the normal process of review that numerous good people stood by it. The EMC voted unanimously to accept the first report, and essentially reject the second. Duty-based ethics had prevailed, although the majority of commissioners would not have formally understood the underlying concept.

A type of example that I like to reference is related to data and its potential for manipulation. One specific example I like to reference is what can be called "the outlying data point." If there is a positive aspect to this example, it is that the misrepresentation of the data can be both unintentional as well as intentional. On the positive side, I reference Virginia Tech professor Geoffrey Vining and his expertise related to linear regression analysis. Vining states that there are potentially three problematic data points: outliers, or points where the observed response does not appear to follow the pattern of the other data; leverage points, which are distant from the other data in terms of the regressors; and influential points, which combine both ideas. For example, in the world on linear regression, the Mahalanobis distance, which is a measure of the distance between a point and a distribution, is a key measure when evaluating problematic data points.

Overall, Vining argues that understanding the impacts of outliers is important in the analysis, and that throwing out data points is not the job of the analyst. Not all researchers or analysts agree. The decision is an individual one and one that is very seldom explained to the decision maker.

I also reference much more simple examples of this issue. A lawsuit in recent years was brought by a whistleblower against Duke University that accused researchers of using false data to win more than $200 million in federal grants. A similar issue occurred at North Carolina State University where researchers eventually were found to have manipulated data to prove a theory and continue to win grants (and promotions). Such occurrences are certainly not restricted to these two universities, but the point of the examples is to represent a larger crisis in research and data analysis.

In 2015, the journal *Science* reported that a study conducted by an international team of more than 300 people concluded that their attempts to reproduce the results in 100 different psychological experiments were mostly unsuccessful. All 100 experiments were published in 2008 in one of three psychology journals: *Psychological Science*, the *Journal of Personality*, and the *Journal of Experimental Psychology: Learning, Memory, and Cognition.*

The take-away is that the results from researchers, academics, scientists, and even public agencies whose mission is to protect the public are all far from perfect. There are few enforceable rules of professional conduct and limited licensure with enforceable penalties under law.

PROFESSIONAL CANONS OF ETHICS

The following list below provides a summary of the standards of ethical practice for a number of licensed professionals and two examples from professional organizations:

- Engineers: Duty-based ethics; included in all state licensing laws; violation can result in license suspension, loss of license and civil penalties; requires placing public health, safety, and welfare and issuing true and objective statements above personal self-interests. Licensed professional engineers cannot turn off their professional standing. The licensee is held to their duty-based ethical standards at all times.

- Licensed medical professionals: Duty-based ethics; included in all state licensing laws; violation can result in license suspension, loss of license and civil penalties; standards and enforcement generally stronger for physicians than nurses; requires placing responsibility to the patient and human health first, next perception of profession, and last, personal self-interests. Licensed physicians cannot turn off their professional standing (licensed physicians are held to the ethical standards at all times).

- Accountants: Duty-based ethics; emphasis on integrity and objectivity; state licensing laws vary in content, more so than with licensed professional engineers and licensed physicians; client confidentiality held in higher regard than with licensed professional engineers, whose primary responsibilities are first for public health, safety, and welfare.

- Licensed attorneys: Attorneys tend to be self-regulated under the law; consequence-based ethics, although some experts opine that legal ethics are indeed duty-based (duty is that justice is served). However, this is not the case in most modern professional codes and state statutes. Attorneys put needs of the clients they represent first.

- American Society for Quality (ASQ) Certified Reliability Engineer (CRE) Ethics: Duty-based ethics; not statutorily binding; failure to comply carry maximum penalty of losing certification and membership in the organization.

- Society of Maintenance and Reliability Professionals (SMRP) Ethics: Pledge to represent their profession ethically and honorably; not statutorily binding; eight key points, including represent their qualifications honestly and their educational achievements and professional affiliations, and provide only those services which they are qualified to perform; and to follow all policies, procedures, guidelines and requirements promulgated by the SMRP Certifying Organization.

SUMMARY THOUGHTS ON ETHICS

- There are three primary forms of ethics—virtue, deontological (duty-based), and consequentialism. Professionals such as licensed professional engineers and licensed physicians subscribe to duty-based ethics which is enforceable by law. Duty-based ethics are most relevant to reliability, risk, and resiliency professionals.

- Default to fair and honest communication. Trust the data and trust the truth.

- Avoid tailoring your communications to one audience. Reliability, risk, and resiliency are by their nature complex and contain uncertainties. Associated decision making is also necessarily complex. Your communications will likely be passed through several different audiences prior to a final decision being made. You only have one opportunity to be credible and trustworthy.

CHAPTER 5

PRACTICAL TOOLS

Our journey thus far has taken us through a foundational understanding of the definitions, jargon, and concepts related to communicating reliability, risk, and resiliency to decision makers. This module focuses on providing a glimpse of converting that foundational understanding into effective, practical applications.

In a workshop setting, we normally spend a meaningful amount of time examining, reviewing, and discussing real-world examples. As with any type of case study, we often learn more from what did not work than from the glossy, positive examples that focus only on what did work. We also learn much from the real world versus applying theories or hard-and-fast rules that should work. All of this really says that time and circumstance are often bigger forces than the robust rules we create concerning the do's and do nots of communication.

As we pivot from foundation to application, I want to make one other note I have presented in my workshops. For all of the arguing against complex graphics generated from commercially available software and stressing that we may not need graphics in every communication, in this section I show some of the commercially available software tools I regularly use. The first reason is that the use of commercially available software is simply a reality on several different levels, including time savings, resource sharing, and quality assurance. The second reason is to provide workshop attendees with some practical examples, tips and pitfalls for some of commercial software tools that I regularly use.

The primary commercial software that I use for communicating reliability, risk, and resiliency to decision makers—and especially risk and uncertainty—is the Palisade DecisionTools Suite. There are others, including products such as those produced by Reliasoft in the field of reliability, and other products like Microsoft Excel and Esri GIS products. However, the Palisade DecisionTools Suite is the best I have personally found for getting decision makers from the data to the decision.

Tools within the Palisade DecisionTools Suite can be found on their website, and this is not intended as an advertisement for their product. The Suite includes tools for Monte Carlo simulation (@RISK), decision trees (PrecisionTree), statistics (StatTools), data patterns (NeuralTools), visualization, mind mapping, and data exploration (BigPicture), and optimization (Evolver and RISKOptimizer). There are many very good competing products that are perhaps as good as the Palisade tools from an analytical perspective. The message is that Palisade has found a way to keep it relatively simple by being committed to the universality of Microsoft Excel and at the same time providing the practical analytical power for end users to get to decisions. I simply prefer the Palisade tools as a single software of preference because, in my opinion, their tools strike the right balance between the analytics and the communication that is necessary for success. That is why they often dominate my collection of real world examples.

The following sections provide a summary of two types of graphics. The first of these is a group of "essentials" for communicating reliability, risk and resiliency to decision makers. Essential is operatively meant to convey "almost always" rather than imply a fixed rule of "always." The second is a group of "graphics to use with caution." Within this group are graphics I frequently use but are often misunderstood; their use depends on their formulation, their delivery, and their relevance.

ESSENTIAL GRAPHICS

This group includes pictures, geospatial depictions, time series charts, tables, tornado diagrams and guiding graphics.

PICTURES

I begin this part of every workshop with the reminder of the adage that "a picture is worth a thousand words." The take-away message is to use pictures early in the communication to make sure that each decision maker knows the physical aspects of the subject of their decisions. From much severe experience,

I have learned all too often that the actual decision makers have never seen their physical facilities, plants, and infrastructure. In other cases, it has been a decade since they have seen them or they have seen an earlier (and different) version of them.

Getting everyone on the same page with a common understanding necessarily requires that the physical picture must be current. There should be some neutral object in the picture to provide the audience with an indication of size and scale of the object of interest. The caption should state the perspective (view looking north toward Church Street) since many decision makers will attempt to mentally frame the situation by the people impacted by their decision. The picture should also relate something relevant and simple to the decision at hand or that is at least interesting to the audience.

My primary tool of choice now is a thermographic camera. These are now very powerful for taking digital pictures of the same object with and without the thermographic image. A side-by-side picture portrays both the natural look of the object and the heat, coolness, and other aspects such as fluid levels, where applicable.   The picture is interesting, provides some new insights even for people familiar with the object, and underscores the communicator's commitment to detail. Flir and Fluke are leading manufacturers of this type of equipment (my personal camera is a Flir E-4). Most cameras are available for under $2,000 and some lower-cost versions are available for use with cellular phones.

GEOSPATIAL DEPICTIONS
Physical locations and physical relationships are also important

for providing understanding. Akin to a picture, many decision makers will also try to orient their perspective and decision impacts based on where and who may be impacted by their decisions. Aerial photographs and maps from commercially available sources such as Google are sufficient.

As discussed in the chapter on Graphical Excellence, aerial photographs should be in black and white (greyscale) or Google-type maps in their natural colors or black and white. Note that most Google-type maps have features provided in light colors or pastels, which, as mentioned in an earlier chapter, prevents the base document from detracting from the primary message. The primary features that are the central point of the message should be shown in darker colors on both the aerial photographs or maps.

Geospatial depictions are important for communicating context and orientation to decision makers.

One or two geospatial depictions are sufficient. The common mistake is to use too many geospatial depictions, which tends to distract the audience from the data and information that is trying to be communicated. The key point here is that some geospatial depiction is needed for decision makers. I have also found that most non-technical specialists cannot effectively understand process diagrams and flow charts without first understanding the geospatial relationships and orientations of many of a given system.

TIME SERIES CHARTS

A time series chart displays observations on the y-axis against equally spaced time intervals on the x-axis. The major advantage of time series plots is that they allow the user to evaluate patterns and behavior in actual or forecast data over time. For financial analysis, monetary values are represented on the y-axis and time (typically in years) on the x-axis.

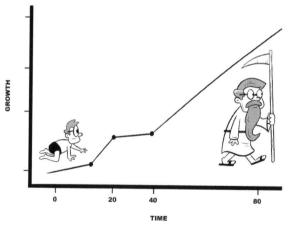

Hindsight can be analogous to 20-20 vision. Our forecasts are at best 80% accurate. Time series graphs are powerful communication tools.

In forecasting, deterministic models use point estimates for each input parameter. From these single-point estimates a single, specific outcome is predicted. Such models are rooted in the theory of determinism, which holds that every outcome event is the result of preceding, or antecedent, events. There are at least four major problems associated with forecasts based on this methodology: the belief that all preceding events can be precisely determined; the failure to account for interdependence among the input events; input events may not have been accurately measured, interpreted, and reported; and the belief that precise models can be developed to capture the relationship of the inputs and output events.

More simply said, the primary problem associated with this approach is its inability to account for uncertainty. A deterministic forecast is an over-simplified approach that reduces its applicability to real-world events. In most cases, this is the reason for the common perception that forecasts are wrong the minute after they are made. An equally important ramification is that a deterministic approach normally drives the forecaster to collect as much data as possible, and with a high level of measurement, in order to avoid being considered wrong with the associated prediction of the future. Many organizations waste tremendous amounts of resources collecting data to potentially improve a deterministic forecast while at the same time not gaining any advantage in the decision-making process.

Probabilistic forecasts use statistical distributions associated with input parameters rather than single-point estimates. According to ISO 31000, the international risk standard, this approach provides a method of taking into account uncertainty on systems in a wide range of situations. It is typically used to evaluate the range of possible outcomes and the relative frequency of the values in that range for quantitative measures of a system such as cost, duration, throughput, demand and similar measures. In short, probabilistic methods allow for the forecaster to better understand uncertainty and risk, and the sensitivity of the relationship of inputs and outputs.

The use of a cone diagram or a box and whisker diagram (sometimes called simply a box plot) is used to depict the results of a probabilistic forecast. The cone diagram resembles a funnel with its narrow end in the present time and its wide end at the end of the forecast period. Normally, a different shade of color is shown to distinguish the central 50th or 90th percentile with the outer extreme ranges. The use of the cone diagram is easier for most decision makers to initially understand.

The box and whisker diagram depicts the same thing as the cone diagram but without the smooth depiction generated by the cone. Its more granular, or choppy, appearance often makes it more difficult for decision makers to initially understand. But in fact, the box and whisker diagram is simply an aerial view of a histogram for each year in which a forecast is made. The median value is represented by a line within the box for each year a forecast is made. The line is synonymous with the point forecast associated with a simple deterministic time series plot. Normally the 25th and 75th quartile are depicted as the box while the most extreme values, typically the 5th and 95th projections, define the ends of the whiskers.

The width of the cone or box and whisker plot allows for a visual representation of the risk associated with the forecast in one graphic. The greater width of the box in any given year implies more uncertainty or potential risk associated with the forecast. Boxes in any given year that are skewed in either direction from the median value also depict the potential for more upside or downside in any given year.

Time series graphs are extremely important for understanding past data trends and future forecasts. For forecasting, probabilistic methods are recommended to provide the decision maker with the best possible understanding of risk and uncertainty associated with the forecast. The cone diagram and the box and whisker diagrams can provide the same information and are equivalent; however, most decision makers and lay people find the smoother depiction of the cone diagram easier to understand initially.

TABLES

Graphs are simply a concise method for depicting large amounts of data. In many cases, graphs are not necessary to provide an understanding of the data. In some case, the use of the wrong type of graph or the poor construction of the graph makes the data more difficult to understand than if the data were simply presented in a table.

A common mistake by technical professionals is to assume their audiences are incapable of understanding the data and information without the use of a graph. This is usually not true. In many cases, the use of a table is preferred to the default use of a graph.

In truth, a table is the source of every graph. In reliability, risk, and resiliency, it is common for decision makers or their associates to request the underlying data that is used to develop the graph. The communicator of reliability, risk, and resiliency data should expect to produce this underlying data.

And the technical credibility of the presentation often depends on producing it. Therefore, the central question is not whether to use a graph for understanding in lieu of producing data; the central question for the decision maker is whether the graph is needed on top of the data in order to produce a more concise understanding of it.

Tables with thick borders, interior lines that separate data, and shadowing of text or the frame are distractions from the data and the message of the table. Therefore, these techniques should generally be avoided. In terms of the data, frequencies should be provided in conjunction with percentages, extreme values should be used in conjunction with a range, and the median value should always be provided with the mean (or average).

TORNADO DIAGRAMS

Tornado diagrams are modified versions of a bar chart. They are a classic tool used to communicate the results of a sensitivity analysis, which can be performed either deterministically or probabilistically. The tornado diagram receives its names by the visual image that is created from wider bars associated with input variables that have more impact on the output being located at the top, while the narrow bars associated with input variables with less impact on the output being shown at the bottom.

It is important to explain a tornado diagram initially to decision makers. I have found the following sequence to be effective: the input variables are shown down the left side of the diagram; the vertical line running through the center of the bars represents central tendency (usually the median); the bars represent the potential range of variation in the vertical line (central tendency) that is the result of that particular input parameter; and therefore, the input parameters at the top of the diagram are much more important than those at the bottom of the diagram.

From practical experience, most decision makers quickly understand the primary takeaway—things at the top of the diagram are much more important than things at the bottom —after one explanation. It becomes a powerful reference graphic in projects with subsequent phases. It will also be a graphic that will become requested as a standard in other types of reliability, risk, and resiliency projects.

Keep in mind two aspects of tornado diagrams that will not necessarily be understood in an initial presentation but will be useful in subsequent presentations. First, the diagram is a powerful quality-assurance and quality-control tool because all of its bars should touch the vertical central value line. If a bar does not touch this line, then that associated input has been programed incorrectly in the model because its variation does not impact the output. In practice, we have found numerous model errors in the work presented by other experts or consultants with this simple test. The tornado diagram therefore provides to the communicator a valuable tool for discussing sensitivity and can also be used as proof of quality assurance if asked by the decision maker.

The second aspect that will require some subsequent follow up with decision makers is the underlying expression of risk and uncertainty as a function of variance in the diagram. The demonstration of variance as mathematically behaving geometrically, actually as a square, rather than arithmetically is extremely important. An input variable whose bar is two times as wide as another input variable is not twice as impactful but rather four times more impactful. However, the general rule for the communicator is that this line of discussion will most likely sidetrack the entire communication and therefore as a rule it should not be used without upfront work with the receivers of the information.

I want to note some practical considerations for presenting tornado diagrams. First, the initial explanation of the tornado diagram can be done in a picturesque manner, meaning that all the input parameters on the left side of the figure do not have to be legible by the audience. It is much more important for decision makers to understand the basic concepts and feel of the "tornado". Subsequent enlargements can drill down into the details of inputs that are most or least important; of course, in the case of enlargements, the inputs parameters should be legible. Second, tornado diagrams by their nature are busy diagrams. Shades of blacks and blue colors are best for the major components. Use sharper colors like greens and reds to highlight key points.

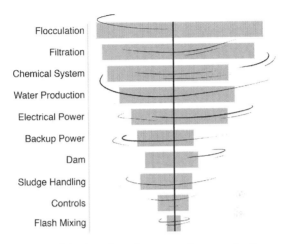

Flocculation
Filtration
Chemical System
Water Production
Electrical Power
Backup Power
Dam
Sludge Handling
Controls
Flash Mixing

The special form of the bar chart known as the tornado diagram is one of the most powerful communication tools. The most impactful system components appear at the top and the least impactful at the bottom. Decision makers can quickly understand where to direct limited resources.

GUIDING GRAPHICS

The analysis and treatment of reliability, risk, and resiliency is iterative in its nature. This requires that the decision maker understand the process and inter-relationships associated with the various components if communication is to be effective. Examples of this are demonstrated in ISO 31000, the international risk standard, and in ISO 55000, the international asset management standard. In practice, the need for a guiding graphic to demonstrate the iterative nature of the work and the progress that is being made is most needed on intermediate and high-complexity projects; however, the value in producing this type of graphic on smaller and less complex projects should not be overlooked.

Management consultants and graphic artists have made a farce out of this type of graphic. Remember that at all times we should avoid chartjunk, visual images that have no meaning, and graphics that distract for the underlying data and information. Avoid at all costs depictions of circular processes, overlapping circles, and boxes and triangles that nest in multiple dimensions without going anywhere. The internet and sales material are full of examples, and they are developed by people without much technical depth and who are usually selling shiny objects.

The guiding graphic that should be associated with the communication of a reliability, risk, or resiliency project should depict a logical flow and at least some primary end point. Above all, decision makers need to know that the process of addressing a complex problem is moving toward some type of finality. The visual feel should be more of a bubble diagram rather that a flow chart primarily because it should produce a feel of warmth and confidence in the process rather than require a deep and time consuming technical understanding by the decision makers. Connecting lines should be minimized since we are looking for general flow and not technical precision. If there is a potential graphic where the colors can be somewhat overdone, it is the guiding graphic.

The guiding graphic should also include points or references to specific milestones where decision makers will be involved. Decision makers, with few exceptions, care little about how the technical "sausage" is made or about the details of how hard the technical analysts are working. They care when and where their input is needed and what kind of decisions will be required at those milestones. In short, the guiding graphic is most about them. The guiding graphic is really not about the data and information either. And it is certainly not about you as the technical professional.

Guiding graphics are important for longer term problems involving complexity and uncertainty. This type of graphic provides decision makers with a quick understanding of the analysis timeline.

GRAPHICS TO USE WITH CAUTION

SCATTER DIAGRAMS

Scatter diagrams are used to describe the relationship of one variable to another. This is most commonly done in two dimensions (one variable of the x-axis and one of the y-axis) although with modern software there is no technical reason why scatter diagrams cannot be shown in three dimensions. Statistical correlation and best fit trend line can be developed from variable relationships and represented on scatter diagrams.

The warning with respect to scatter diagrams is that correlation does not imply causation. Many decision makers and lay people are led down the wrong path by believing that one variable causes another when the two variables are highly correlated. This is not the case. Regardless of the amount of effort and analytics that are used to unwind a potential misconception, once the seed is sown in the human mind that one variable causes another when there is high correlation it is extremely difficult to reverse.

I do use scatter diagrams on a regular basis in my communication. However, I frequently remind myself that they do not capture variation over time, typically do not capture magnitude, and correlation between two variables does not imply causation. I use scatter diagrams in communication with caution.

The scatter diagram is a powerful tool for technical analysis. It can create confusion as a primary communication tool.

MATRICES

Matrices are a form of scatter plots. Common matrices include Heat Maps in the field of risk and PICK diagrams in the field of Six Sigma. Perhaps the most famous matrix in the general business world is the Boston Consulting Group's Growth-Share Matrix (1977). This matrix shaped the marketing industry by providing a tool to help evaluate opportunities by the relationship of market growth rate and a firm's relative market share. The four sectors were called Stars, Question-Marks, Cash Cows, and Dogs. Many other matrix examples can be found in general business, especially related to portfolio classification, evaluation and optimization.

The same issues related to scatter plots apply to matrices. With matrices used for risk analysis, issues related to time and magnitude are less obvious and not appreciated by many practitioners as well as decision makers. An additional consideration with respect to risk matrices is the potential for risk mitigation to reduce certain risks more than others – in other words, not all risks that are considered as "high" are equally impactful once an equal amount of risk treatment is applied to the initial classification. Another consideration related to risk matrices concerns scales and measurement theory and the fact that what is plotted in a risk matrix is a compilation of subjective judgement rather than quantitative data.

A fine point in communication of risk matrices that compares the likelihood and consequences of failure is that different technical sectors have plotted the data differently. In the finance and insurance sectors, consequence (usually in dollars) has been historically plotted on the y-axis and the likelihood of failure on the x-axis. The plotting of monetary values on the y-axis is consistent with a time series graph. It also has certain intuitive appeal to financial and insurance professionals because the plot of monetary values on the y-axis provides a visually intuitive financial cut-off level. Engineers plot the data in an opposite manner, with consequence shown on the x-axis and the likelihood on the y-axis. This is likely the by-product of the engineer's action level being aimed at controlling/understanding the likelihood of failure. Both produce the same answer, and neither is wrong. However, the format of the graph may create confusion or misunderstanding for the decision maker depending on their core field of expertise. Credibility and distraction are the primary issues for the communicator to address.

I use risk matrices regularly because in practice they are easy for decision makers to understand. As a form of technical best practice, I believe that they are very flawed. My personal preference would be for most manuals of practice simply to not use them as standard default graphics. At a minimum, their technical limitations should be more fully discussed in manuals of practice. Similar to scatter plots, I use risk matrices with caution and on a limited basis.

HISTOGRAMS

Histograms are used to show frequency distributions related to how often different values in a data set fall into a certain range. They are especially helpful in understanding the dispersion (range) and central tendency of a data set. For technical analysis, they are perhaps the most important graphic for quickly and effectively understanding basic data properties.

However, decision makers and lay people have limited intuitive understanding of histograms, and the communicator needs time to explain them simply. I have numerous visuals that I use to explain histograms, including dice, beads, coins, and wood models. Nevertheless, their lack of intuitive appeal makes them have a high potential for distracting decision makers away from the core message of the communication. I prefer not to use them in my primary communication; however, my general advice is to use them with much caution.

Many technical professionals are bothered by this position of not using histograms as a form of primary communication since frequencies and probabilities are core foundations of the analysis of reliability, risk, and resiliency. However, the issue that must be addressed is whether this line of detailed discussion is necessary to encourage decision makers to make an irrevocable allocation of resources. In most cases, it is not. If decision makers are curious about the many analytical details, there will be ample opportunity to address their questions. However, in practice few questions on these details are asked by decision makers.

If a discussion of frequency distributions and the related histograms are important, I have learned from my colleague Adam Sharpe that the most effective way to communicate them is by using comparative analysis. Displaying multiple histograms on the same graphic allows for the audience to focus on the factors that create differences rather on trying to fully understand the aspects of an individual histogram and subsequently to use it in application. Highlighting differences has intuitive appeal.

PIE CHARTS

Pie charts add nothing that cannot otherwise be expressed in a table or stacked bar chart. Analogous to the common misunderstanding of probability, they are easily misunderstood unless the associated frequencies are attached. They are likewise simply a snapshot in time. Pie charts can tell us how much, but cannot help provide any understanding of why.

My experience is that pie charts have a higher potential for adding misunderstanding, or worse, nothing new, than they do for adding something to the decision-making process. There are simply better graphics that can be used. Therefore, I use pie charts sparingly and cautiously.

SUMMARY THOUGHTS ON PRACTICAL TOOLS

- Essential graphics include pictures, geospatial depictions, time series charts, tables, tornado diagrams and guiding graphics to help decision makers understand the analysis process. For example, the adage "a picture is worth a thousand words" is universally applicable to reliability, risk and resiliency communications. Essential is operatively meant to convey "almost always" rather than imply a fixed rule of "always."

- Scatter diagrams, matrices, histograms, and pie charts are included in a group of "graphics to use with caution" when communicating with decision makers. I frequently use these graphics, but they can be often misunderstood, over-simplified, or cause confusion in trying to explain them. Their use is highly dependent on their formulation, their delivery, and their relevance.

- In practice, graphics need not be simple nor complex. A best practice is to keep one key message per graphic. Each graphic should show something unique and meaningful from the data. If not, do not use the graphic.

CHAPTER 6

NOISE

Communication is the exchange of information or ideas from one person to another. It has three elements: a communicator (or sender), a message with a means of transmitting information or ideas, and a receiver. Effective communication occurs only if the receiver understands the exact information or idea the sender intends to transmit (US Army, 1984).

FORMS OF COMMUNICATION

I normally use the simple example that Dan Vallero and I have used in a paper and panels on which we have both served. The picture is a radio tower propagating signals from the top. In workshops, I typically add a cartoon for extra effect.

Successful communication is analogous to the signal-to-noise ratio (SNR) in electronics. Breakdown between the signal generator and the receiver occurs during message delivery. In electrical and communications engineering, noise is commonly defined as unwanted energy that degrades the quality of signals and data. A breakdown of signal, including too much noise, is not the fault of the receiver. It is the fault, and the responsibility, of the sender.

Noise is analogous to signal-to-noise ratio in electronics.

One way to look at potential noise is through considering that information is either communicated perceptively or interpretively (Vallero et al, 2007). Perceptual communication occurs through the five senses—visual, audible, olfactory (smell), tactile (touch), and taste. Much of the general public communicates perceptively. The consequences are probably obvious in terms of communicating reliability and risk; sometimes instincts are good but often the evaluation is subject to bias. Communication related to resiliency is even more difficult because the general public and decision makers often have little previous experience with the rare events. This lack of previous experience occurs both in the planning for a potentially rare event and in its aftermath.

In terms of communicating reliability, risk, and resilience, there is a balancing act between playing to the senses—and instinct—and overplaying to the senses which trigger biases. The dangerous result of playing to the senses is that you are introducing noise that distracts from the signal. The core message is always the same; let the data speak for itself and present it the way you would want it presented to you. I'll discuss more on this topic in subsequent chapters.

Interpretative communication is the other form of communication. Two major subdivisions within interpretative communication are symbolic communication and verbal/conceptual communication. For example, engineers communicate interpretatively and symbolically, using mathematics and diagrams. Other professionals, such as lawyers and policy makers, also communicate interpretively but typically chose to communicate verbally.

For many technical professionals, noise is defined as too much slick talk, too many philosophical concepts, and too few numbers. On the other hand, noise to lawyers and policy makers consists of too many statistics, too many numeric models, and too much detail on any one perspective.

A recent example that I use is from my service on the Environmental Management Commission (EMC). As noted early in the chapter related to ethics, the EMC is the state's official environmental rule-making body and regularly deals with topics related to air quality, water quality, and waste management. Public involvement is a major component in rule making. EMC commissioners rotate serving as hearing officers who basically serve as facilitators of public meetings and compile summary comments to the full commission. Public comments are a major basis for the environmental rules that are adopted by the EMC.

In this example, the state's stormwater rules were being reviewed and re-adopted in accordance with 2013 regulatory reform legislation, and also a fast-track permitting rule was being adopted in accordance with different legislation from the same 2013 legislative session. In addition to my experience with stormwater design and construction, I was familiar with the rules through service on a Technical Review Workgroup in 2013 and 2014 (consisting of engineers and DEQ technical staff) and service on a Minimum Design Criteria (MDC) team in 2014 and 2015 (consisting of a variety of technical professionals). When the revised rules went to public hearing in 2016, I was named as the hearing officer for all three meetings scheduled to gather public input from across North Carolina.

The end of the story is predictable in terms of communication. Over the course of three years of comprehensive debate among a variety of technical professionals, the vast majority of the communication was both interpretative and symbolic. Much of the resolution had been resolved via technical compromise associated with numbers, models, graphs, and technical research. In the public arena, the comments were much more perceptive. A meaningful amount of comments were interpretative and conceptual in nature, particularly those from attorneys and environmental stakeholders. A minority of comments from the public comment period were interpretative and symbolic, and the ones which were of this type were from technical professionals attempting to correct the numbers.

I want to note two interesting items. The first was a comment from the lead person from DEQ, a licensed professional engineer, to me as the hearing officer. Through this experience, she realized that engineers and non-engineers communicate in much different ways. In many ways, the comments from engineers were much easier to deal with in the rules because they were usually related to action levels supported by numbers. The other types of comments were equally important but required much more effort to incorporate into the definitive language of the rules.

The other item of interest was whether the presentation of the material was different in the technical and the non-technical (public) settings, thereby driving the different types of comments and debate. The simple answer was no. The duty-based ethics of the engineers who were involved made sure that the same presentations were given to all. In fact, portions of public presentation were too technical to be understood quickly and completely by laypeople, so additional communication approaches were used to support the base presentation.

COMMUNICATION FORUMS AS A SOURCE OF NOISE

Our communication approach and response to communication forums is also a source of noise. The four forums that I discuss are team meetings, public speaking, communication with the media, and communication with elected officials.

INTERNAL AND EXTERNAL TEAM MEETINGS

One communication forum is the frequent team meetings (internal and external) where reliability, risk, and resiliency professionals commonly present. This type of forum is normally one where the technical professional has been called upon to present technical information. This communication forum is the primary forum to which this book is directed.

There are two subsets within this communication forum. The first is the technical meetings where decision makers are not involved. I argue that the same techniques discussed in this book and the workshops are equally applicable to technical team meetings. If for no other reason, it will make the meetings more efficient and effective. However, there is no doubt that the technical details are much more important because that is the purpose of the meeting – to work out the technical issues. This is the extent to which I talk about this aspect.

The second subset within this forum is the actual communication with the decision maker(s). And of course, this is the topic of this book and the workshops in their entirety. The key point I leave here is that communicating reliability, risk, resiliency to decision makers requires a fundamentally different approach than communications related to public speaking, with the media, and with elected officials.

Reliability, risk, and resiliency professionals should be concerned with communicating in a truthful, ethical manner that is effective. In this context, effectiveness is getting the decision maker to commit resources. The communicator should not be concerned with the final decision over the type and amount of resources. Effectiveness is the primary point of the communication.

Persuasion is the next conceptual step beyond effectiveness. It involves bringing another party to a mutually beneficial solution which, in this case, means a mutually beneficial allocation of resources. Manipulation is another level beyond effectiveness and persuasion. Being persuasive is to some degree part of being effective; however, it is very secondary to the reliability, risk, and resiliency professional. The decision, and a vested interest in the outcome, is not the concern of the technical professional.

Manipulation involves bringing another party to a solution that is beneficial only to the communicator's viewpoint, which is neither ethical nor the purpose of a reliability, risk, or resiliency professional. Again, the reliability, risk, and resiliency professional does not own the decision and therefore has no need to manipulate the communication for one purpose or another.

Three other types of communication forums that I discuss below are public speaking, interaction with the media, and responses to elected officials. In these three cases, specific needs of the receiving audience may drive the presenter of reliability, risk, and resiliency information to customize the content. In the next chapter, I will describe the level of customization that may be required to account for the forum and the particular members of the audience. However, such customization should be considered a secondary objective. It will also create some level of noise that potentially distracts from the data and the associated information that is being provided.

PUBLIC SPEAKING

In my definition, public speaking involves providing 30- to 60-minute presentations to community-based organizations, stakeholder groups, or professional societies. For reliability, risk, and resilience professionals, this type of public speaking may take the topical form of information about a new product or technology, a response to a major threat, or a summary of actions taken on a specific subject. Lunch or dinner presentations, key note speeches, and many conference presentations fall into this category.

The most important distinction between a presentation on reliability, risk, and resiliency is that it should be focused on the data and balanced in its overall content. Public speaking is about the entertainment value. Said another way, technical presentations are data-oriented while public speaking is speaker-oriented.

Conference presentations serve as a good example of the potential differences. My colleagues and I often ask ourselves whether the purpose of such presentations is to convey new information and approaches or simply to provide a forum for collecting continuing education credits in some nice venue that can dually serve as a form of vacation. Purists will say the former while realists will say the latter. In most cases, the highest ranked conference sessions are usually provided by realists. Their presentations normally have more pictures than words, stories to make you laugh, and few process diagrams. High ranked conference sessions are usually energetically given in a shorter time period than provided in the agenda so that everyone has some extra time to socialize, get a cup of coffee, or take a bathroom break before the next session starts.

One way to look at public speaking is to view it as an appeal to the sensory stimulations of humanity. As discussed earlier, the frame is more related to sensory communication—see, hear, touch, taste, smell—than to interpretive communication that is practiced by most engineers and scientists. The trappings of sensory communication are noise.

This is not to say that communication-related reliability, risk, and resiliency should not be entertaining. As the old saying goes "every time you open your mouth you enter show business." For example, graphical excellence is always a must. But what also is true is that you only get one shot at credibility and trustworthiness in technical communication. The take-away is be conscious that anything done for the showmanship of public speaking is also a source of noise that distracts from the signal that you are transmitting.

WHAT THE MEDIA WANTS

Interaction with the media is also a different communication forum. I have had the primary responsibility for dealing with local and state level media at different points throughout my career. And I have had a number of formal media trainings, from a variety of sources, during the past thirty years. Additionally, I developed my own training for technical professionals based on my training and much severe experience. My starting and ending point to technical professionals is to leave communication with the media to the media communication experts.

In the context of communicating reliability, risk, and resiliency, consider any attempts to communicate with the media as simply another forum for creating noise. This is in part the result of the media's need to communicate with the public in a sensory manner; however, it is also in part based on the communication being interpretative. That interpretation is left to someone else—the reporter or the editor. The minute you say words or show data you have lost control over the interpretation.

Individuals in the media come in all shapes and sizes, just like any other profession. There are several major types of media: newspapers, magazines, radio, television, and social media (in simplification, on-line reporters or bloggers). Each have subcategories, such as local newspapers versus major market newspapers or general magazines versus technical journals. Each major type and subtype of media are looking for slightly different information or take different perspectives.

When the media calls, they are seeking information. Timely information is considered news. Most reporters have limited technical knowledge, so they are often looking for an expert—and preferably an expert with a strong opinion with a good quote (at least a good quote for their perspective on the story). The media also may genuinely want a good story, an eye witness account, or a good visual image. The main point is that they want information and preferably in a manner that helps them (the media) interpret events as they see them. The unintended consequences can be scary.

I use an example in my workshops from a recent libel case in North Carolina. The quick summary is that the State Bureau of Investigation (SBI) had been providing bad testimony in criminal trials based on marginal testing techniques. The *Raleigh News & Observer (N&O)* was accused and found guilty of libel related to six statements the paper published regarding an SBI firearms examiner (technical professional).

The *N&O* reported that executive editor John Drescher told a Wake County in his testimony that investigative journalism is the newspaper's core mission but noted that such articles are "extremely expensive" to put together and "don't sell papers."

An outside forensics expert had emailed the *N&O* reporter just 11 days before the story ran, explaining he did not perform firearms analysis and that his criticism of forensic analysis was not directed at any specific case. In other words, the technical expert had been quoted out of context or the quote was misinterpreted.

The *N&O* reported the defense attorney asked executive editor John Drescher, "Did anybody on the (investigative) team come to you and say, 'John, we have this pretty harsh quote, and the email saying he doesn't do firearms examination'?" When Drescher said no, the defense attorney asked, "So, no 'Houston we have a problem'?" A Wake County jury later awarded a $1.5 million verdict against the *N&O* for libel.

Most of the cases where technical professionals have been misunderstood or misquoted by the media do not go to trial. However, the occurrence is frequent where technical professionals are misunderstood or misquoted and usually with negative consequences on the technical professional. Dealing with the media is both an art and a science that requires formal training and much severe experience. Again, my advice to reliability, risk, and resiliency professionals is to leave it to the experts. In our communications, consider it a source of noise.

WHAT ELECTED OFFICIALS WANT

Interaction with the elected officials is a fourth communication forum. Like my experience with media communications, I have also had the primary responsibility for dealing with local, state, and federal elected officials throughout my career. I have also served as a registered lobbyist at the state level, served as county political party chairman, served as a trustee and chairman of a political action committee (PAC), and served as a political appointee at the local and state level. I have had a number of formal trainings in dealing with elected officials and, similar to media training, I also have developed my own training for technical professionals.

My starting and ending point to technical professionals is to leave communication with elected officials to the formally trained communication experts. With that said, communicating with elected officials is very similar with dealing with boards of directors. In many cases, they may be one and the same. Avoiding commentary with elected officials or board of directors may not be as simple as minimizing public speaking opportunities or leaving the media to someone else.

My advice is the same. The reliability, risk, and resiliency professional should consider communicating with elected officials as another forum for creating noise. In large part, elected officials embody the general population, so communication is at least in part sensory based. However, interpretation is equally important to the political operative—after all, politics is simply one way for a group of people to bring another group of people to their same understanding. Persuasion through interpretation is king.

It is important to remember that elected officials seek only a handful of things: to do something good and be part of the solution; to get re-elected or re-appointed; to get things done that are important to them; to avoid making people mad at them if they do not care much about the issue; and to not look bad or stupid to the public, their colleagues, or the media.

The handful of important things can also be said about most members of boards of directors.

Somewhere in my experience I was given three rules for dealing with elected officials. These have always proven effective and are simple: be accurate, be brief, and tell me something new. Elected officials do not want to be embarrassed, do not have much time, and use sensory communication with their constituents, which is the reason that all politicians like a story.

The real danger for creating noise when dealing with elected officials is that reliability, risk, and resiliency are not easily broken into meaningful and universal sound bites which can be used to support or persuade one person to another's position. In this case, the noise may not be generated from the initial communication but from the cascading communication that follows. Intentionally or unintentionally, the credibility and trustworthiness of the reliability, risk, and resiliency professional will suffer, and the intended signal will not be interpreted correctly by the receiver.

COLOR AS A SOURCE OF NOISE

Color is just one source of noise that may distract from a clear communication signal. It is a subset of visual image, which can include items ranging from page layout, font type, and visual image of the presenter. Visual image, written (or verbal) image and vocal image are the three forms of image most frequently described by communication experts.

Written (or verbal) image, includes items like grammar, word choice, sentence structure and abbreviations. My college writing handbook was *Hodges' Harbrace College Handbook*, which is still the oldest but has slipped in popularity. I use several other writing references, including *The Gregg Reference Manual*. My favorite is now *Booher's Rules of Business Grammar* because of its simplicity in targeting the top 101 business writing mistakes. Regardless of the exact source, I recommend for everyone to have at least one printed writing handbook for definitive reference in addition to the many electronic ones that are also available. While written image is important, I have little new to say on the subject that cannot be found in the many good references that are already commercially available.

Vocal image includes things like pronunciation, tone, pitch, and pace. It is the third image that can be a source of noise that distract from a clear communication signal. Like visual image and written image, there are numerous good resources that are commercially available. Recordings are more powerful than books in this subject area. At a minimum, I recommend taping practice speeches and communications to hear how you sound. Take advantage of any formal training opportunities to get in front of a microphone or camera with experienced trainers.

For reliability, risk, and resiliency professionals, the data should be able to stand on its own. The data are the purpose for your report or presentation. Sensory communication is indeed important as a secondary consideration, but interpretative communication is the primary reason for you being there.

With that said, most workshop audiences like the discussion of color. I think the source is dual in nature—on one hand it is a primary consideration in graphical excellence and associated chartjunk, and on the other hand, many corporate communications professionals obsess (and lecture technical professionals) regarding its use. Certainly, there must be value in its discussion.

I recommend keeping it simple when it comes to font colors. Black and dark blue are typically associated with finality, credibility, and calmness. Darker shades of green of a similar effect as darker shades of blue. Red and yellow as font colors are usually distracting, may be difficult to see, and as a font color produce noise. Orange can also be distracting for similar reasons. Purple seems to be recently popular but is usually interpreted as whimsical and therefore not well suited for instilling authority in a technical report.

As stated in the chapter on graphical excellence, grey scale and pastels are good colors for background images on most graphics. Areas of emphasis can be effectively shown in using blacks or blues. In this case, reds, yellows, and oranges can be used for strategic emphasis.

In workshops, I normally provide a traditional overview of colors that are common in the communications and branding industry.

Red has several different contextual associations. On one hand, it is the color of financial loss, blood, fire, and danger. On the other hand, because of its association with danger and heat, it carries connotations of passion, boldness, and energy. Red is best used with discretion because it has powerful meanings on each extreme. However, it is a powerful color for adding emphasis (it can simultaneously appeal to both energy and cautiousness).

Blue is a universally preferred color. In branding, it is a favorite color for conveying reliability, dependability and trustworthiness. It is generally appreciated for its calming and harmonious qualities, including its association in nature with the sea and sky. However, blue is also associated with the emotional feeling of being in a state of sadness or depression, or "blue." Blue is a solid color in almost any application, and the decisions around it are normally in the selection of shade rather that whether to use it or not.

Green is associated with nature, finance, and wealth. Although the naturalist and the financier may be at different ends of the political spectrum, green has positive meaning for both. Traditional logic is that lighter shades of green indicate growth and vitality, while darker greens represent prestige and wealth. Green also is considered to promote discussion and interaction since it generates positive emotions.

Yellow is an attention-grabbing color. It is the most common color for highlighters. Yellow can communicate energy and happiness, and has also been found to stimulate mental activity. It is also used universally for caution lights, marking hazards such as steps or slippery surfaces, and even the color of crime tape used by police. It produces too much glare for use in text and as a background color; however, it can be used for its strongest purpose—to highlight and draw attention to key points.

Orange is a blend of the positive attributes of red and yellow. It traditionally has been associated with vitality, freshness, and youth. Similar to yellow and red, it is good for highlighting and drawing attention to key points. It is not typically associated with business, seriousness, or finality.

Purple is rarely found as a natural color in nature, which is probably why it has been traditionally associated with royalty, spirituality, and mystery. Behavioral psychologists have found it to be a low arousal color. Lighter shades of purple such as lavender are usually more popular with women than with men. Because purple is seen as a unique, mysterious, and uncommon color, it has the effect of being associated with things that are either optional or whimsical. Purple may have its place as a highlight color or secondary color, but should not be used by reliability, risk, and resiliency professionals as a "go to," base color.

In workshops, I typically state, "hopefully at this point this discussion is interesting." For most it usually is. For some participants, it is distracting. And that is the point. My experience is that the data, the information, and the message can—and should—stand on its on in the communication of reliability, risk, and resiliency. Sure, the attention to fine points like color are important from a secondary perspective. The data can carry the day regardless of the color, but the color cannot carry the day without the data. Visual image, verbal image, and vocal image are all sources of noise if overplayed.

PATTERNS OF COMMUNICATION AS A SOURCE OF NOISE

Communication patterns describe the structures by which information flows in a group. These structures should be reflective of a group decision making processes, the desired efficiency of the group, and the responsibilities of each member in the group. The effectiveness of communication also has a big impact on the satisfaction of the group members. Communication patterns can also be a source of noise.

Harold J. Leavitt is given credit for the development of the most common categorization of patterns of communication. His 1950 research is somewhat limited in scope, but his five simplistic categorizations are sufficient for reliability, risk, and resiliency professionals to understand this topic and appreciate its potential to create noise.

Leavitt's five structures are chain, circle, wheel, Y, and network. I briefly discuss each in the following paragraphs.

Chain, or line, describes a one-way communication pattern. Usually, this is in the form of top-to-bottom or bottom-to-top of an organization. All members cannot communicate with the leader or primary decision maker in this structure. Feedback can be distorted. The "chain of command" is one manner to visualize the chain, or line, communication pattern.

In the circle communication pattern, a person sends a message or information that can be received by all members of the group. However, the leader or decision maker can only communicate directly with others who are peers in the hierarchy or direct reports. It resembles the chain communication pattern in this respect, and again feedback can be distorted since communication must flow up the chain of command through superiors. In an organization structure, lower-level staff appear to not have a role in the decision making.

In the wheel, or star, pattern the leader or decision maker is at the center of all communication. In this structure, all members of the team can communicate with the leader and vice versa. However, members are not connected to one another and, in some cases, may not know the other team members (groups), who may be advising the team leader, exist. The wheel is often preferred to the line or circle communication pattern because it is quicker and prone to less distortion (or noise).

The Y communication pattern is like the chain or circle. The concept of sub-groups is introduced in this communication pattern and each sub-group has a leader who serves as the primary point of communication. Sub-groups may have different patterns of communication within their respective sub-groups, but the effect is that the sub-group leader still is the point person in a top-to-bottom or bottom-to-top communication pattern. Sub-groups only communicate with each other through the sub-group leaders.

The network communication pattern allows all member of a team or decision process to communicate with anyone else based on their information needs and decision requirements. The network communication pattern has some major advantages related to feedback loops, less potential for distortion from organizational hierarchies, and time of communication. However, it can also produce inefficiencies since more people are usually included in the communication chain than may be strictly needed. There are often rules and guidance provided to keep network communication patterns efficient.

Social media is a form of a network communication pattern. It doesn't take much imagination to visualize the inefficiencies created by receiving too many messages. Likewise, we simply have to look to the way Donald Trump used social media in the presidential election of 2016 to understand how a network communication pattern can be very effective in reducing the potential noise created by other patterns of communication.

Patterns of communication exist in business settings just as they do in family settings. They can be sources of noise and misunderstanding.

Patterns of communication are yet another form of communication that can produce noise. Like communication forums and communication image (visual, verbal, and vocal), patterns of communication should be treated as sources of noise.

SUMMARY THOUGHTS ON NOISE

- Communication is the exchange of information. It consists of a sender, a message with a form of transmission, and a receiver. The responsibility for effective communication of a message belongs to the sender.

- Communication can be viewed from the perspective of a signal-to-noise ratio. It is the sender's responsibility to eliminate as much noise as possible so that the signal can be properly received. This means that the burden of effective communication falls on reliability, risk, and resiliency professionals and not their superiors, board members, or decision makers.

- Four sources of noise are described. The first source relates to how people process information. The general public usually uses a sensory form on interpretation, lawyers and policy makers use an interpretive-concept form, and engineers, scientists, and other technical specialists use an interpretative-symbolic form. Too much or too little of any of the three forms tends to create noise.

- A second source of noise is the forums of communications. Public speaking, interaction with the media, and interaction with elected officials provide three sources of potential noise since the communication objectives of each forum are different from the primary communication objectives of most reliability, risk, and resiliency professionals.

- A third source of noise is attempting to appeal too much to sensory issues. Color was used as an example but is really just a subset of a larger set of sensory communication issues related to visual image, verbal image, and vocal image. Noise can be created by the reliability, risk, and resiliency professional trying to focus on these from a primary objective standpoint. Let the data speak for itself.

- The fourth source noise that was discussed is the pattern of communication. In this case, the pattern of communication (line, circle, wheel, Y, and network) also introduces the potential for noise given the hierarchies through which the communication must pass. There are no perfect patterns of communication; however, the reliability, risk, and resiliency professional should acknowledge their potential for creating noise and incorporate measures to minimize them.

- Effective communication is the responsibility of the sender, not the receiver. By its very nature, reliability, risk, and resiliency are complex topics based on symbolic analysis which introduces noise to many receivers. Keeping it simple is the best way to reduce noise, while keeping in mind that making it too simple creates noise for some receivers. The balance is in letting the data speak for itself, and presenting it in a balanced and ethical manner regardless of the immediate audience to which you are trying to appeal.

CHAPTER 7

THE AUDIENCE

The previous section described noise as the part of a message that distracts from the signal that the sender is trying to communicate. I described sources of noise for awareness and avoidance by the sender (you). However, an additional issue is the degree to which noise is a function of the receiver's ability to process the communications.

In the formal discipline of communications, the receiver is the audience. The standard approach is that understanding the audience is usually the first step in developing a communication plan and, in turn, communicating effectively. The corollary is that all communications should be tailored to the receiver. So why is the audience discussed now instead of sooner?

Hopefully that answer is well understood by now. Reliability, risk, and resilience communication should stand on the data first and foremost. The communication should be balanced, truthful, and ethical. If it is effective, then a quality decision (or an allocation of resources) will be made. Communication of reliability, risk, and resilience information should be most about effectiveness of supporting the decision maker to making a quality decision. It should be less about persuasion, advocacy, or manipulation. Communicating reliability, risk, and resiliency is different than selling a car, getting someone to buy your house, or being elected to public office.

Nonetheless, all communication shares a component called an audience (the receiver). It is both not practical and naive to assume that communication is not impacted by the specific audience. There is always a degree to which communication can be enhanced by understanding the audience.

With this acknowledgement, I also provide a statement of caution. Decisions that are made with respect to reliability, risk, and resilience are seldom made in one setting; therefore, the special effects that are used to motivate one audience may be the same ones that will not work, or worse, have a negative effect, on another audience. The adage of "painting oneself in a corner" applies.

Less obvious is determining exactly who the audience is. Reliability, risk, and resiliency are by nature complex. The related decision making is neither impulsive nor short-term. Standard decision making theory states that a decision is made by one individual. But is this necessarily true for complex decisions that require longer periods of time to make? And if so, what is the role of the group on the one decision maker? Even as an area of secondary priority, understanding the audience related to reliability, risk, and resilience is complex.

PERSONALITY PROFILES
(OVERVIEW AT INDIVIDUAL LEVEL)

I provide four tools for understanding behavior, communication, and decision making. These tools are Myers-Brigg, DISC, handwriting analysis, and neuro-linguistic programming (NLP). The former two are quite common in corporate settings. Additionally, other models exist.

MYERS-BRIGGS

The theory behind the use of four quadrants to better understand an individual's personality style can be traced to Empedocles in 444 B.C. He attributed personal behavior to four external, environmental factors, or elements, Earth, Air, Fire, and Water. Hippocrates around 400 B.C. attributed individual behavior to four internal factors (or humors), which he described as Choleric, Sanguine, Phlegmatic, and Melancholic. A person's most dominant humor was believed to determine their personality.

Swedish psychologist Carl Gustav Jung, a contemporary of Sigmund Freud, re-examined these four quadrants and associated types of behavior. He concluded that personality styles are indeed internal and that the difference in personality style is related to the way we think and process information. Several of the other methods used to assess individual personality styles can be traced to students and followers of Jung.

Jung's book *Psychological Types* was published in 1921 in German and translated in 1923 into English. Following the line of development of Myers-Briggs, American Katherine

Cook Briggs began her own independent study around 1918 based primarily on Jung's early work. She, too, subscribed to four quadrants to describe basic personality style. Briggs four quadrants were thoughtful, spontaneous, executive, and social; Jung's were thinking, feeling, sensation, and intuition. Briggs' first two papers were "Meet Yourself Using the Personality Paint Box" (1926) and "Up from Barbarism" (1928) and were predictably rooted in Jung's work.

Briggs' daughter, Isabel Briggs Myers, joined her mother in her research. She later went to work for Edward Hay, who would start one of the first management consulting firms specializing in human resources. The *Briggs-Myers Type Indicator Handbook* was published in 1944 and in 1956 changed its name to Myers–Briggs Type Indicator, or MBTI. MBTI was adopted and further developed by a number of different testing and psychological organizations. It now has several subsequent editions and it is perhaps the most common individual personality assessment tool used by businesses.

An interesting fact is that neither Briggs, Myers, or Hay were formally educated in the field of psychology.

Anyone who has participated in an MBTI evaluation will remember the four-letter description of personality style. For example, an "ESTJ" personality style is described as extraversion (E), sensing (S), thinking (T), and judgment (J). An "INFP" personality style is described as introversion (I), intuition (N), feeling (F), and perception (P). In workshops, participants normally have a good time comparing their similarities and differences to others they know. Most participants also are exposed to the more serious side of how personality style impacts the way an individual audience member perceives and interprets information related to communications, sales, negotiation, and decision making.

Jung believed that people are born with preferred ways of perceiving and deciding. The MBTI sorts some of these psychological differences into four opposite pairs with a resulting 16 possible psychological types. None of these types are "better" or "worse". However, all people have preferences that impact their comfort and willingness to accept information. A corporate trainer once told me that anyone can do any job for a certain period of time, but that people will only be happy with a job over the long term that matches their personality type. This comment from the corporate trainer was a page from Jung, and he was exactly right.

The finer points of Jung and the Myers-Brigg Indicator Test are beyond the scope of my workshops. I do like to reference MBTI and provide a little background since it is both popular in business environments and relevant to gaining a better understanding of the audience. However, my own preference is another tool which I have found to be just as effective, more intuitive, and requires less time to administer.

DISC

William Moulton Marston (1893–1947) was a Harvard-educated American psychologist. His pen name was Charles Moulton. A unique character by almost any standard, Marston's greatest achievements were providing the basis for an individual personality style assessment tool, developing the systolic blood pressure test which is still part of the modern polygraph (lie detector) test, and creating the fictional character Wonder Woman. The blood pressure test was greatly influenced by his wife, Eliza Marston, and Wonder Woman by his longtime partner, Olive Byrne. The Marston's and Byrne had a longtime polyamorous relationship, including William Marston fathering children by both women and Eliza naming her daughter, Olive, after Olive Byrne. Marston is also known as a feminist theorist in certain circles.

Marston published *Emotions of Normal People* in 1928, just after the publication of Carl Jung's book *Psychological Types* and about the same time as the papers published by Elizabeth Briggs. Similar to the official development of MBTI in the period immediately after World War II, Marston's theory was not developed into an assessment tool until 1956 by industrial psychologist Walter Vernon Clarke. It was branded as DISC after Marston's four dominate personality styles (dominance, inducement, submission, and compliance). A number of different entities have since produced their own versions of the DISC assessment.

Marston's approach is appealing for its relative simplicity and remaining generally consistent to the very earliest forms of personality style description. In addition to some of the descriptive names that I have encountered and are shown below, I have been part of different assessments tools that equated DISC to birds, wild animals, and even biblical characters.

D:Choleric, Dominance, Power, Driver, Competing, Controlling-Taking, Direct

I:Sanguine, Inducement, Expressive, Collaborative, Adaptive-Dealing, Influencing, Spirited

S:Phlegmatic, Submission, Denial, Analytic, Avoiding, Conserving-Holding, Supportive, Systematic

C:Melancholy, Compliant, Suppression, Accommodating, Supporting-Giving, Cautious, Considerate

Marston believed that personality styles came from people's sense of self as well as their environment. Like Jung, he believed that people could adapt their personality to some degree to their environment, but in the long- term their behavior was driven by their natural, internal norms. Later in this chapter, I will discuss more on the use of Marston's approach for the practical evaluation of the audience.

HANDWRITING ANALYSIS

Handwriting as an indicator of personality dates back 2000 years. Aristotle noticed the correlation between handwriting and personality. The Chinese independently observed a connection between character and writing. In 1622, the first book was published in Europe describing the analysis of character through the study of handwriting.

In the late 1800s, Abbe Michon wrote several books on the subject and coined the term graphology to describe the practice of handwriting analysis. One of his students, Jean Crepieux-Jamin (also a dental surgeon), classified the many features of graphology into a comprehensive system. Both Michon and Crepieux-Jamin were of the French school of thought. Adolf Henze, a contemporary of Michon, is credited with founding the German tradition.   Ludwig Klages, a German philosopher of the same period, applied gestalt theory to graphology in the early 1900s and is credited with formally applying scientific, proof-based methods to it. Max Pulver, an associate of Carl Jung, used psychoanalysis in the interpretation of graphology in the early part of the eighteenth century and was an antagonist to Klages. Ania Teillard, who also worked closely with Carl Jung for several decades, applied Jung's approaches to the theory of graphology. More recently, psychologist Alfred Binet, who invented the first practical I.Q. test, confirmed that character traits are reflected in handwriting.

Like fingerprints, handwriting is very specific to an individual. This is one reason why law enforcement agencies, including the US Federal Bureau of Investigation, use handwriting analysis as one of numerous investigative tools.

The Trait Stroke method and the Gestalt method are the two major schools of thought related to handwriting analysis. The Trait Stroke method, which has origins associated with the French, holds that a specific stroke formation (like a wide "o") reveals a certain personality trait. There are more than 120 different strokes that indicate a separate personality trait, so in combination they provide a powerful profile of an individual. As handwriting expert Bart Baggett says, "a stroke is a stroke, no matter where you find it." Or for that matter how you find it. Research has found that stroke characteristics are similar even for people who have lost their primary writing hand, and must use the alternate hand, their feet, or their mouth to write.

The Gestalt method has German roots and developed at approximately the same time as the Trait Stroke method in France. The Gestalt method focuses primarily on how the words are written on the page. More simply, handwriting can be viewed like to a painting. Wide borders (white space) tells you something about a person's personality that is different than limited borders or borders that favor the bottom, top, or either side of the page. The spacing between words tells us something too, as does the vertical spaces between lines. Similar to the Trait Stroke method, there are many finite details that make up the analysis.

In truth, both methods really take us to the same place, and many analysts use a combination of both. My formal training is in the Trait Stroke method. I have self-studied the Gestalt method and have found it very useful. Both methods, and handwriting analysis in general, provide useful ways to understand people better and in an abbreviated amount of time. Their best use is in helping to understand what types of questions to ask someone, to get a better understanding of their reactions to a given situation, and to use in building rapport between the sender and receiver of information.

NLP

I mention neuro-linguistic program (NLP) as another tool for understanding an individual's behavior and in doing so it helps to build rapport. Some of the concepts related to NLP provide some power around the edges of understanding an individual, but tools like DISC, MBTI, and handwriting analysis are better core tools.

NLP has many critics. It has fallen far enough out of favor with the scientific community that there is limited published research on the subject in the past two decades. Many academics and psychologists hold to the position that NLP has since been discredited scientifically in the field of psychotherapy. However, it still has a meaningful commercial following and is used as a tool in some business management seminars and workshops. The internet is full of pro-NLP and anti-NLP debates.

Proponents of NLP claim it is the study of what works in thinking, language, behavior, and even decision making. It considers individuals to be the product of both their conscious and their unconscious. One way to look at this is consistent with the adage "you know a person by what they do, not what they say." By studying a person's basic thought patterns—visual, auditory, and kinesthetic (touch/feelings)—you can understand their natural traits and behaviors. These traits and behaviors play out in the words people use, their eye movements, and their body language.

The study of thinking patterns pre-dated NLP, but thinking patterns are a core concept of NLP that helps reliability, risk, and resiliency professionals better understand an audience. For example, individuals who use terms such as "sounds good," "struck a note," "sound the alarm," and "say to yourself" likely favor auditory thinking patterns. A presentation or report that incorporates some form of sound may be more favorably received than one that does not. On the other hand, individuals who use a dominant amount of terms such as "feels good," "hit me like a ton of bricks," "food for thought," "building blocks," and "reach new heights" likely favor kinesthetic thinking patterns. A presentation or report that includes some form of touchable or tangible objects may be more favorably received than one that does not.

I pause here to address a potential point of contradiction. In Chapter 2, I pointed out that Clark (2010) has debunked the myth that visual thinkers require visual images to learn. If this is true, then the logical question is related to why we need to understand thinking patterns in our communication. The answer is that not all communication is related to learning. In NLP, thinking patterns are a key aspect of building rapport and relationships. Clark did not say that graphics or visuals do not enhance communications; she only points out that visual thinkers do not require images to learn effectively.

Addressing audience preferences related to visual, auditory, and kinesthetic thinking patterns can have an impact, especially in an age where there are many options by which the sender can transmit messages.

Anyone who has watched the movie "Meet the Fockers" appreciates Robert De Niro's character studying Ben Stiller's eyes to see their movement to gain an understanding of his thinking patterns, his character, and ultimately whether he is lying.

The study of eye movements is more than one hundred years old and predates NLP. However, the study of eye movements is a key concept in NLP for understanding an individual's subconscious self. For example, individuals who look up, either upward left or upward right, while answering a question are using visual cues to recall images. Individuals who look in the horizontal plane are usually using auditory cues to remember what was previously said. Recalling feelings and inner reflection is usually indicated by looking down.

When an individual is looking away, in any direction, he or she is accessing information. There is no reason for further discussion until the person makes eye contact again. But there is some thought or reflection to what is being said. If a person never looks away, or maybe never looks back, then this is likely either a sign of lack of interest or a fixed position on the topic.

There is much more that can be determined from the study of eye movement, including indications as to whether an individual may be lying or whether they are stimulated (usually indicated by eye pupil dilation). My general focus is to make technical professionals aware that these types of concepts exist, but I do not teach or claim to be an expert with these items.

The aspects of NLP which are beneficial for reliability, risk, and resiliency professionals is in building rapport through understanding a person's unconscious preferences. NLP advocates typically argue that decisions are not made primarily based on the technical information but rather based on the relationship between the provider of the information and the decision maker.

Because NLP is another method for understanding the audience, I often reference some of its material in longer workshops. Other tools are better in my opinion, but some insights can be gained from an awareness of some of the key concepts like thinking patterns and eye movements.

RELEVANCE TO INDIVIDUAL COMMUNICATION AND DECISION MAKING THROUGH DISC

I prefer to use Marston's four-quadrant approach as my primary tool for discussing individual personality styles and their impact on communication and decision making. The main reasons for using this tool is that it is relatively straightforward to understand, relatively simple to provide examples, and over time I have found it to be the most well received. DISC also provides an appreciation of the influence of personality styles for individuals that make up any audience but touches on the topic only at a higher level—after all, the key message here is to let the truthful and ethical presentation of the data stand on its own and not to get wrapped up in too much detail that may inadvertently create noise.

Marston's four-quadrant approach can be viewed in two dimensions. The top and bottom quadrants in most DISC assessment tools represent whether or not a person is externally or internally focused. The left and right sides represent whether a person is more oriented to action or to people.

The Dominance and Inducement personality types are externally focused and the Submission and Compliance personality types are internally focused. The Dominance and Compliance personality types are task focused while the Inducement and Submission personality types are more people focused. Individuals typically have two primary personality traits with one being more dominant than the other. There are many different combinations that define a personality style, and in turn shape a person's communication and decision making preferences.

People who have primarily a Dominance personality style have a fundamental need for a challenge and for making decisions. Under stress, they tend to take control and address the problem (usually at the expense of the human element). Because they are focused externally and on taking action, long presentations, fluffy details, and making a strong recommendation will create trouble when working with the strong D personality type. Fortunately, there is a relatively low percentage of strong D's in the general population.

People who are primarily an Inducement personality type have a fundamental need for recognition and interaction. Under stress, they will attack other people because they are concerned that controversies or problems will make them look bad to others. They care a lot about image and who else is doing what they may be doing. The best way to lose in a presentation with a strong I is to leave the door open for someone to question their role in a negative way; the way to win with an I is to show them that the solutions you are proposing are "tried and true" and will reflect positively on them.

One practical question relates to how best to play our odds in terms of connecting with individual preferences. First, the numbers show that just over half of the general population are more people oriented than task oriented. Therefore, including in your communication or references how the issue will impact people will produce a good potential for positive effect. Second, the numbers also show that approximately two-thirds of the population are internally focused. Including in your communication references to proven processes, quality control, and proven methodologies will produce a good potential for positive effect.

The ultimate question is how does an individual's personality style impact their decision making. The caveat is that individual decision making goes far beyond personality style. However, personality style does play a role.

Understanding personality styles and using approaches that minimize noise will improve communication effectiveness with decision makers.

Years ago, I heard the analogy between personality types and shooting a gun. I do not remember the original source, but the analogy is a good one. In decision making, D types tend to be ready, shoot, aim; in other words, the need for action leads to quick decisions and unintended consequences. For the people-oriented I types, the sequence is more like ready, ready (whose watching?), aim, aim (whose watching?), shoot (who else shot?). For the S types, the sequence is ready, ready, ready.... And for the task oriented but internally focused C types, the sequence is ready, aim, aim, aim....

The combination of personality types complicates and also enhances an individual's decision making. Other strong factors are individual biases and the group effects. So individual personality is an interesting piece of decision making, but only a piece.

INDIVIDUAL BIASES AND DECISION MAKING

Rules of thumb, or heuristics, are used to make decisions in the face of uncertainty. This is both a good and a bad thing. On the good side, humans are required to make many decisions under uncertainty every day. Rules of thumb help us make most of these decisions quickly and effectively. On the bad side, some decisions are more strategic in nature and require more detailed analysis. Rules of thumb in these cases oversimplify the problem and the solution, and in turn lead to poor-quality decisions. Rules of thumb can create biases and are at the same time the product of biases.

In an earlier chapter I cited the simple example I use in workshops related to confronting a tiger. If I walk out of a building and see a tiger, my rule of thumb (or my heuristic) is simply "run." This assures my survival and the survival of my family or my co-workers or whoever may be with me. It also ignores the fact that this may be a pet tiger, an old tiger, a toothless tiger, or even a tiger on a chain. I simply run because my "rule" is a tiger is a potentially bad thing, and this means run.

There is another type of tiger that I see more regularly because my son goes to Auburn University. His name is the mascot Aubie, and he is a human dressed in a tiger suit (by the way, he is most often wearing a jersey with a 1). My rule of thumb in this situation is to say "Aubie!" and extend my hand for a high five. The vast majority of the time it works well. But I am uncertain of the person who is in the suit, and there is a possibility that this person could do harm to me.

Psychologists and behavioral scientists describe a prevailing model of two types of thought, or cognitive, processes. These have traditionally been called intuition and reason. More formally they are now referenced as dual-process cognition models, and more generically referenced by technical specialists as System 1 and System 2 thinking based on research performed by Stanovich and West.

Each works equally well when applied in the right situations. In fact, experience developed by applying the logic, and often numeric thinking, required for System 2 can eventually lead to better intuition that forms the basis of System 1. Ideally, System 1 and System 2 are complimentary in the sense that System 1 provides quicker, more intuitive judgments while System 2 allows us to review, correct, and even override System 1 judgements. In the real world, things like root cause analysis and continual improvement fall under the realm of System 2. The caveat to this ideal, complementary relationship between System 1 and System 2 thinking, is that it seldom happens naturally.

Reliability, risk, and resiliency are by their nature complex and typically based on data and analysis. They fall into the realm of System 2. The ramification is that there will be a problem if the technical professional is communicating in System 2 ("reason") but the decision maker believes the problem requires System 1 ("intuition"). The point is the same in the opposite direction if the technical professional "dumbs down" the communication because there is a belief that the decision maker views the problem as System 1 when in fact the decision maker has classified the problem in the more complex realm of System 2. Decision making is fluid, based on situation and timing, and therefore determining with great accuracy whether an individual is classifying a problem as more simple or more complex in nature is difficult. It is very difficult in a group setting.

Decisions require the allocation of resources. So whether intuition associated with System 1 or whether reasoning associated with System 2 is dominating an individual's thinking processes, many decisions are based on the beliefs of the decision maker concerning the likelihood that uncertain future events will happen. In planning, this may involve supply and demand for certain products by consumers. In engineering, operations, and maintenance, this may involve the likelihood of failure of a certain piece of equipment or physical system. In finance, this often involves the time value of money and whether interest rates are going up or down. In regulatory affairs and public policy, it usually involves who will win the next election.

These beliefs, shaped to a large degree by our past experiences, form the bases of our biases. Not all rules of thumb, or heuristics, are biases, but biases frequently shape the rules of thumb we apply in decision making. The effect of the bias on decision making is what is important since all individuals are prone to biases. Clarifying and accounting for individual biases is the primary reason that formal decision quality processes, like business case evaluations, are used to support big and/or complex decisions. The circular issue is often associated with getting an individual to understand their bias of oversimplifying a big and/or complex problem.

Daniel Kahneman and Amos Tversky are pioneers in the field of behavioral psychology and judgements under uncertainty. Their research, papers, and books are worthy of study. I usually recommend their work with Paul Slovic captured in *Judgement Under Uncertainty: Heuristics and Biases* and Kahneman's *Thinking: Fast and Slow* as must reads. Of course, there are many other exceptional experts in this field of study, some of which are collaborators with Kahneman and Tversky and some without association.

In my workshops, I usually provide a list of 18 common biases that I have developed from my experience and study in this area. There is nothing magical about the number 18 other than my own opinion and ability to provide practical, tangible examples of each. In their 1974 article *Judgment under Uncertainty: Heuristics and Biases*, which led to their 1982 book of the same name, Kahneman and Tversky described three major categories of heuristics that lead to biases. As quoted directly from the article (Kahneman and Tversky, 1974), these are "(i) representativeness, which is usually employed when people are asked to judge the probability that an object or event A belongs to class or process B; (ii) availability of instances or scenarios, which is often employed when people are asked to assess the frequency of a class or the plausibility of a particular development; and (iii) adjustment from an anchor, which is usually employed in numerical prediction when a relevant value is available. These heuristics are highly economical and usually effective, but they lead to systematic and predictable errors. A better understanding of these heuristics and of the biases to which they lead could improve judgments and decisions in situations of uncertainty."

GROUP BEHAVIOR

At this point in the chapter, there should be some appreciation for the complexities associated with understanding the individuals that make up the audience. We now turn to issues related to the audience acting as a group.

Fundamentally all decisions are made by an individual. However, we only have to look to the "mob effect" to understand that most individuals are greatly affected by their associations with a group. In many cases, individuals make decisions to comply with group expectations that they would not make if alone.

Kahneman and Tversky are academic leaders in the field related to individual decision making. Their peers in the field of group decision making are Howard Raiffa, Herbert Simon and James March. Raiffa is known in the field of group decision making and specifically for reaching consensus and making decisions with groups of independent stakeholders. He is also one of the forefathers of multi-criteria analysis. Simon and March are renowned for their work with group behavior and decision making in large organizations, namely government organizations but also large corporations. I also add Colin Talbot to this list for his work related to correlating organizational structure and performance measurement in the public domain.

However, the one example that I use, at a minimum, in my workshops is from a less known psychologist, Robert C

Fried, from his 1976 book *Performance in American Bureaucracy*. I like his concept on individuals acting in a group because it is simple and practical. Fried argues that decision making in small, private-sector organizations is primarily for efficiency, and this serves as the common thread that drives the decision making by the group. By efficiency, he means financial profits, and I expand it to say that the foundation for profitability is happy customers and quality products/services. Fried argues that this is true in large public and large private organizations too, but there are also additional factors that drive group decision making.

The first additional factor is described as a Democratic Ethic, which in my terms is what makes the boss happy. In large organizations, mid- and upper mid-level managers often have different values than the leadership at the organization's top. Mid- and upper mid-level managers often have greater influence on an individual's long-term career than the people at the very top. The second is described as a Legal Ethic, or more generally the influence of elected officials and community leaders but which are outside the direct chain of command within the organization. Keeping these people happy is often just as important as keeping the current agency top leadership happy. And third, Fried uses the term Liberalism not in the political sense but rather to describe that especially all public organizations must be respectful and open to all. Decisions are often made to make the organization reflective of all, even if such a decision is not otherwise in the organization's best interest. Fried's opinion is that decisions made by large organizations, and especially public organizations, are impacted by several factors not associated with individual preferences, organizational efficiency or the formal chain of command.

Using this one model by Fried, the reliability, risk, and resiliency professional gains an appreciation for the exponential complexity of understanding the audience as the dimension of the audience migrates from the individual to the group setting. On one hand communication is usually more effective the more you understand about the audience, and on the other hand the complexities created by attempting to tailor too much to the audience often leads to noise and other negative consequences.

Should our objective be to isolate the primary decision maker and tailor our communication solely in that direction? The answer is no. From a practical standpoint, the majority of organizations require passing information through the chain of command and, as such, the information must be passed through several internal audiences. An even more fundamental reason is that we want complex problems to be deliberated by a group.

Paul Schoemaker, an expert is strategic management and decision making, has observed that there are motives beyond compelling social reasons for group cooperation. He believes that individuals simply have too many information processing biases and physical limitations to solve certain types of problems. We simply cannot avoid the reality that involving a cross-functional group of experts is the best way to ensure quality decision making. Maybe less obviously, reinforcement through group involvement is also very important to most individual decision makers when coping with uncertainty and complexity.

Individuals simply have too many information processing biases and physical limitations to solve certain types of problems without input from others.

The role of experts in decision making has been roundly debated for a long time. Paul Meehl's 1954 classic, *"Clinical versus Statistical Prediction,"* was the first blow to the tradition that experts were much better evaluators and judges than models based on numbers. Robyn Dawes' 1976 paper, "The Robust Beauty of Imperfect Linear Models" concluded that neither experts nor robust numeric analysis was much better than disaggregation and weighting of decision criteria. This idea was another form of what Howard Raiffa had championed a decade earlier, which was that getting to decisions on large and complex problems required a multi-criteria approach that brought stakeholders to consensus. More recently, Phil Tetlock has evaluated exactly what makes an "expert" an expert and in many cases the prediction of experts, especially political and policy experts, is no better than what the rest of us can predict.

The point related to experts is complimentary, not contradictory, to Schoemaker and the prevailing logic on group participation in decision making. Whether it's data, or expert opinion, or the simple models created from stakeholders, no one approach is substantially and consistently better at predicting the future and making good decisions than the other. We need all three. It is an important realization for communicating reliability, risk, and resiliency to decision makers.

The potential negative impacts of the group as an audience should also be considered. One example from Herbert Simon is related to group and organizational loyalty. Simon observes that the values and objectives that guide individual decisions in organizations are largely the organizational values and objectives. It does not really matter whether these are expressed or implied. Initially, these are usually imposed by the exercise of authority on the individual. However, the values and objectives gradually become internalized and are incorporated into the psychology and attitudes of the individual participant. In other words, the individual becomes conditioned and in turn acquires loyalty to the organization that automatically guarantees that his decisions will be consistent with the organization values and objectives.

This conditioned loyalty becomes a hurdle for effective communications and decision making related to complex problems with uncertainty. On one hand, it assures protection of the organization from outside forces. On the other hand, it is one reason why change is hard and good decisions related to complex problems is so difficult.

Another example of the negative impacts of the group as an audience can be seen through what Kahneman and Tversky call the Planning Fallacy. The Planning Fallacy is used to describe the plans, forecasts, and decisions made by groups that are unrealistically close to best-case scenarios and that can be improved by reviewing statistics of similar cases. The simple example in practice is despite the technical recommendations provided to the group, a project has a budget, schedule, and quality of deliverables that exceeds what has ever been previously delivered. In other words, the group ignores the data and expert opinion and talks themselves into an overly optimistic and unachievable plan. It happens routinely in all types of group settings.

The takeaway from these two examples of the negative impacts of the group on decision making is the recognition that the group effect is a powerful inside force that is difficult to overcome in communication. One approach is to roll with it rather than to fight it. Kahneman described the "outside view" as something that is needed to neutralize the negative effects of the group. Reliability, risk, and resiliency professionals are usually better served as being the source of the outside view rather than the inside agent of change against the internal view. This is especially true since the role of reliability, risk, and resiliency professionals is usually not as the ultimate decision maker.

Carl Spetzler credits his colleagues at Strategic Decisions Group with introducing a best practice for achieving clarity and alignment in making significant organizational decisions. The Dialogue Decision Process, or DDP, "moves decision makers toward a quality decision through a dialog with the decision team around specified staged deliverables" (Spetzler, 2007). As a student of Spetzler, I have used DDP numerous times in recent years. I have found it to be the best model for working with a team in making strategic decisions with complexity and uncertainty.

SUMMARY THOUGHTS ON THE AUDIENCE

- Standard guidance on communications usually recommends that first you should understand your audience. In most cases, there is limited guidance or tips in how this should be accomplished. This chapter has attempted to provide some guidance, tips, and appreciation for the complexities of the issue. Understanding the audience is both a science and an art.

- A better understanding of the audience will in most cases create more effective communication; however, it is both an important and secondary objective for the reliability and risk professional.

- The data and information should stand on its own. The primary role of the reliability, risk, and resiliency professional is presenting it in the most concise, truthful and ethical manner possible.

- The role of the reliability, risk, and resiliency professional is not as the primary decision maker and should not be to persuade or manipulate the decision maker in one direction or another.

- The primary focus of the communication for complex problems should be more focused on minimizing noise and thus assuring that the receiver—the audience, the decision maker—receives as clear of a technical signal as possible.

CHAPTER 8

RESILIENCY AND RARE EVENTS

Rare events are classified as those having a low likelihood of occurring and a high consequence when they actually occur. They surprise us. We want to get back to normal as quickly as possible. Messages sent and received prior to the event are often discounted because the event is rare; communication after the event is difficult because rare events are usually not a single event but instead a series of related events.

As discussed earlier, resiliency is the ability to return to the original form or state after being bent, compressed, or stretched. An alternate definition of resiliency is the ability to recover or return to the desired state readily following the application from some form of stress such as illness, depression, or adversity. The people impacted by rare events expect truthful and accurate communication regarding how resilient their systems are, or simply how quickly their systems will return to their original form so that their life can get back to normal.

Resiliency and risk associated with rare events present a special communication challenge to reliability, risk, and resiliency professionals. The challenge before and after the rare event is both physical and psychological.

THE NATURE OF RARE EVENTS

There are three indicators that an event may be a rare event. From the interpretive perspective, we can say that an event with a low likelihood of occurrence is one that should be considered rare if indeed it were to occur. For many events, we can look backward into history and determine the probability of an event's past occurrence and feel reasonably sure that it will not happen with regularity in the future. The earth being hit by a meteor and a tsunami on the US east coast are good examples. And those rare events where the likelihood can be historically evaluated and also have the most consequences are the ones that our minds are naturally the most concerned. Most people are not concerned, or even notice, rare occurrences that have little impact on their daily lives.

A historical approach to rare events probably gets us only halfway to identifying rare events for planning purposes. Of course, the question related to estimating the likelihood of any event from a historical perspective is the quality of the data. In many cases, historical data are poor or do not exist. Often the estimation of likelihood of occurrence must be a subjective one, or one based on perceptions and biases. And of course, there are the events that we have never seen or experienced— the black swans—where there are no data and where there is no basis for subjective opinion. The impacts of future sea level rise (which I argue should be considered as a rare event) is an example.

Another indicator of a potentially rare event is the way humans treat a potential event once it has been identified. By their nature, rare events are infrequent in their occurrence. People who are in power often believe the event will never happen, and if it does, they have the control to deal with such events just like any other emergency. The source of this perspective is often two-fold. First, many biases are in play, especially the availability bias and the over-confidence bias. Second, people in power often have to make resource allocation decisions, which normally require limited resources to be allocated against more common events. Most people in power want to believe their prioritization and allocation decisions will not cause harm or unanticipated consequences.

On the other hand, people who are most likely negatively impacted by a potential rare event will want some protection. They are not in power or control, and therefore do not have the same biases or feeling of ownership of larger resource allocation decisions. The need for protection usually plays out in the form of a demand for more legislative or regulatory protection against the event. Outside stakeholders also become involved if there is a sense of disproportional or unequal impact and join the lobbying efforts of those that are most likely affected but who are not in power.

If people in power are minimizing the potential impacts from an event's occurrence and if the people who are most affected by it place great emphasis on its potential, you have a good candidate for classification as rare event for planning purposes. An event like the impacts of sea level rise may be debated to classify it as a rare event under the historical likelihood test, but certainly it qualifies under this test.

The third indicator of a rare event that I use is simply to back into the classification by understanding what causes crisis. Again, a rare event is a low-likelihood, high-consequence event that by its nature causes a need to return to the normal state as quickly as possible. This is one way to look at what can be considered as a crisis.

I was exposed to Ian Mitroff in the late 1990s as I was working with clients to prepare for the Year 2000 (Y2K) event and then again as I led my own business out of the dot.com bust of the early 2000s. From Mitroff's book *Managing Crises Before They Happen*, I take away seven categories of crisis: Economic, Informational, Physical (key plants and facilities), Human Resource, Reputational, Psychopathic Acts, and Natural Disasters. Each category has several different types of crisis. For example, natural disasters can include floods, earthquakes, hurricanes, droughts, wild fires, and others. Mitroff's point is that fundamentally, we need to have crisis management processes, including communication, in place for each category of event since the response to different items in each category will be similar. To prepare for multiple event types in each category before preparing for at least one event in each category leaves us vulnerable in our response.

In many ways, rare events can be considered analogous to crises from a planning and communications standpoint.

I like to use this approach for backing into what should be considered a rare event because being able to snap back from a crisis is really the point of being resilient. Brainstorming on what causes a crisis is one way I work with clients to identify risks, especially rare events, and to develop risk mitigation, planning, and communications. This approach also helps in defining associated reliability statements and the basis of designs that incorporate resiliency.

PHYSICAL CONSIDERATIONS OF RARE EVENTS

Resiliency is a byproduct of unreliability. In the absence of failure, there would not be a need to return to an original condition.

Reliability defines an item's intended function, performance standards, assumed operating period, assumed environmental conditions, and time period over which the item will meet its intended functions. The basis of design for any system is based on the reliability statement. Operations and maintenance can worsen the inherent reliability of a design but can never improve it. More simply put, you have to have the right tool for the job; inherent reliability of the tool, or system, is established in design and the reliability statement. Risk and the need to be resilient are a function of unreliability.

As I mentioned in an earlier section, availability and dependability are terms that are used conjointly with reliability to further clarify time periods and, in turn, expectations related to the percentage of time over which an item is considered to deliver its intended functions.

Two common approaches to systems design are prevention-based and risk-based. Prevention-based approaches attempt to reduce the chance of failure to highly significant levels (say one chance in a million or one chance in a billion). Prevention-based approaches are relatively expensive, especially when it comes to initial capital expenditures. Risk-based approaches are relatively less expensive because they accept a higher chance of failure (say one chance in a thousand or one chance in one hundred thousand).

An example of a prevention-based approach could be a reinforced concrete dome, comparable to a structure that protects the nuclear reactor at a nuclear power plant, that would be constructed to protect a military compound in a war zone. An alternative approach could be to construct a system of tall, reinforced concrete walls, which would sufficiently protect the compound from horizontal-type attacks but which would always have some risk exposures for vertical-based attacks such as mortars or bombs from airplanes. This latter, risk-based solution would be more affordable to construct. However, it would carry a higher degree of risk and uncertainty for its occupants as well as a higher need for resiliency planning.

One major issue with risk-based approaches is the way which we assess the potential for the likelihood of failure, especially rare events which include the unknown-unknowns. It is also a problem with prevention-based approaches, although we have at least accounted to a higher degree for the known rare events. However, every system has some risk of failure, and resiliency intends to bring us back to the original level of reliability.

I often smile when I hear resiliency professionals discuss the need for designs that incorporate resiliency as if it were a new concept. The truth is that we have been designing for resiliency for a long time in various forms. First, nothing precludes us from incorporating resiliency concepts into reliability statements and, in turn, our basis of design. In fact, concepts such as Design for Availability or Design for Maintainability have done just that for over 50 years.

Second, in practical terms, reliability, availability, and dependability are judged by the public to mean the amount of time that transpires until basic systems are restored to their original state. Time without service is the enemy, and one reason everyone from the power companies to the airlines race the clock when things go wrong. We have understood the relationships of reliability and resiliency for a long time. The question with respect to physical systems is not whether we understand the relationships, but whether how much we want to pay to reduce the predominately subjective likelihood of failure—and make systems more resilient—on the front end or on the back end.

In terms of resiliency and the special case of rare events, the issue is only one part physical. Risk is defined as the effect of uncertainty on objectives. Effect is dependent on expectations, and both effect and objectives depend on where you sit. For the person in power who is not impacted directly by the event but must be accountable to short-term capital investment decisions, the upfront risk usually is perceived to be small. To the person who does not have the power to make investment decisions but who may be directly or disproportionately impacted by the rare event, the risk is often perceived to be large. For rare events, the psychological consideration related to expectations and perceptions are usually much more important that the physical considerations.

PSYCHOLOGICAL CONSIDERATIONS OF RARE EVENTS

Daniel Kahneman, Amos Tversky, and Paul Slovic are among a number of pioneers that have explored the issues surrounding the psychology of rare events. At a minimum, I like to reference two examples from them when discussing the psychology of rare events.

The first example is related to Kahneman and Tversky. Kahneman and Tversky developed Prospect Theory as an alternative for the traditional decision-making theory, Utility Theory. Utility Theory maintains that decisions are made based on the decision maker using criteria (importance) weighting that matches the likelihood of the criteria occurring. Prospect Theory uses decision weights and likelihoods, too, but in Prospect Theory, the two are not the same. In Prospect Theory, humans tend to be influenced by their expectations and emotions and make decisions under uncertainty that may be mathematically illogical.

People tend to overweight unlikely events and overestimate the probability their occurrence. Again, this is especially true if the decision maker feels out of control and potentially directly or disproportionately impacted by the event. The more vivid the description of a rare event is made to be, the more it is either feared or desired (like winning the lottery). This is one reason that the availability bias is a factor when planning for the potential rare events like tornados or terrorism if such events have recently occurred and were heavily publicized. It is also the reason why fear is a popular line of messaging in political advertisement—the more negative expectations that are established in the public's mind, the more a given rare event will be over-weighted and the likelihood be of the potential occurrence will be overestimated.

A parallel concept is what Kahneman describes as the fourfold pattern. The fourfold pattern relates to the combination of low and high likelihood related to high and low consequences (gains or losses). In the quadrant of a low likelihood event but one that has a potential large gain, the decision maker is likely to not take a short-term compromise or result. One example may be a politician who believes that a rare event is unlikely so the value to his voters is to be fiscally conservative and underprepare for the event. As long as the event does not occur, the gain was worth the risk.

On the other hand, if the likelihood of the event is small but the potential loss is high and immediate, then the effect is opposite. The same politician in this situation becomes risk adverse and may be willing to spend tax payer money to protect the people from impending loss. Thus, moving perceptions from the expectation of a large gain to a large loss is equally important than changing the likelihood of its occurrence.

Prospect Theory is grounded in the perspective of expectations, or choices, from description. An alternate way to look at rare events is based on choices from experience. Because some people have experienced different outcomes from the same rare event, they are often affected differently in terms of how they assign probabilities and weightings in making their decisions. In other words, a person who has actually experienced a hurricane without major losses will likely underweight the likelihood or consequences of the event than someone who has fared differently in their experience or even someone who only is basing their choices on description.

Its sounds simple in retrospect, but it is important to remember that decision making is about looking forward and not looking backward. Kahneman and Tversky maintain that for rare events, the perceptions matter more than the actual probabilities.

A second example is from Paul Slovic. Slovic also has done extensive work in risk and rare events since co-authoring

Judgements under Uncertainty: Heuristics and Biases with Kahneman and Tversky. One observation that he and other researchers have found is what Slovic has called "denominator neglect." Denominator neglect is the observation that people pay attention to only the numerator in a ratio if the numerator supports the position that they have mentally staked out. The classic example involves two urns, one with 1 red marble out of 10 total marbles and the other with 8 red marbles out of 100 total marbles. Participants are asked which urn provides a better chance of picking a red marble on one draw. The answer of course is the first (ten percent versus eight percent) and most people chose it. However, a relatively high number (more than 30 percent) usually pick the second because it gives them a higher number of winning marbles. This is an example of optimism over-riding probability, similar to the way Prospect Theory describes how most people make decisions under uncertainty.

A real-world example for me comes from a major project where my client had little experience in the subject area but liked the idea of being considered an innovator in his industry. We performed probabilistic analysis using Monte Carlo simulations to help provide an idea of the chances of financial success. When asked by the client whether the deal looked like it was financially viable over a certain period of time, I responded that our evaluation had indicated that there was an approximate 90% chance that it would be profitable in the desired time period. When he asked how did I know the approximate chance of success, I replied that we ran 1,000 simulations, with 900 simulations having a positive result and 100 simulations having a negative result. He nearly fell out of his chair.

He regained his composure as the team continued to discussed the analysis and then said to me that he understood that I had said that the project was viable and had a 90 percent chance of success. I replied that what he said was correct. He then turned red, looked miffed, and said "how can you say that – it failed 100 times, 100 times!"

His director of engineering and most of my technical team looked confused. Apparently, the chief executive did not understand what probability meant. The meeting dismissed shortly thereafter with an agreement that everyone was to regroup the next day. The next day the chief executive told me that he understood what probability of 90 percent meant but that he still could not get over 100 failed simulations. Whether he understood probability or not, I came to realize overnight that he feared the big commitment of resources. Because he feared failure more than valued the potential for success, he looked at only the number of failures (100) and ignored the denominator effect (1000). It was illogical to the technical professionals, but our decision maker was unconsciously struck by the denominator effect.

In this project example, the denominator effect worked to support the negative where in the classic urn example it supports the positive. In both cases, the psychological aspects impact how the receiver understands the information. This is usually more pronounced in the case of rare events where the receiver has little experience and must rely on description.

Slovic and Weber stated in their paper, *Perspective of Risk Posed by Extreme Events*, the common belief held by many social scientists that "Risk does not exist 'out there', independent of our minds and cultures, waiting to be measured." Risk is more subjective than objective. It is a concept invented by human beings to help understand and cope with the dangers and uncertainties of life. Again, according to Slovic and Weber (2002), "Subjective judgments are made at every stage of the assessment process, from the initial structuring of a risk problem to deciding which endpoints or consequences to include in the analysis, identifying and estimating exposures, choosing dose-response relationships, and so on."

I like Slovic' s single sentence from his 1999 paper, *Trust, Emotion, Sex, Politics, and Science: Surveying the Risk-Assessment Battlefield*, that "Danger is real but risk is socially constructed." This is particularly true for rare events in terms of measures to plan, prepare and protect against the event and then again in the response to the event's occurrence. Communication is important on the front side of a rare event but is equally, or more important depending on the system's design and reliability, in the aftermath of the event.

PLANNING FOR A RARE EVENT

The standard risk framework identified by ISO 31000-2009 provides a solid basis for rare event planning. Common to all risks, the risk framework contains five primary components: establishing the context, risk identification, risk analysis, risk evaluation, and risk treatment. Communication and monitoring & review are two parallel components that touch all the previously referenced components.

One aspect of planning that is especially important for rare events is establishing the situational context. Reliability statements are an important component of establishing the context for rare events since they establish the boundary conditions around how reliable, or unreliable, a system is designed to be. In other words, they help to establish the context of beyond which point our resiliency planning and communication should be based. Or conversely, once failure related to a rare event occurs, to what point are we trying bounce back and what are the forms of communication that are needed as we bounce back.

The previously-referenced three indicators—historical review, sources of concern or lack of concern, and working backward from categories of crisis—are relevant ways to identify rare events beyond other methods of risk identification. Scenario planning is a powerful tool for identifying rare events because the method is a structured tool for stretching normal analysis and can also identify embedded biases.

Risk analysis for rare events can take several forms. One of the few quantitative methods is by using probabilistic analysis in the form of Monte Carlo simulations. In this context, I briefly discuss risk analysis for rare events and the related special communication.

Monte Carlo analysis is used to quantify risks of all types. In broad terms, risks can be considered in terms of those normally to be expected by a moderately experienced person, through what can be referred to as common sense, or developed normatively through group consensus. These risks fall into two categories: high-likelihood, low-consequence and high-likelihood, high-consequence. Analytically, the expectations related to these type risks can be modeled with continuous probability distributions. For strategic decisions, this type of analysis should be performed at a minimum.

By their nature of being rare, the impact of rare events usually wildly exceeds expectations. In practice, you know rare events after they occurred when you hear comments like "I could never imagine this," "I have never seen anything like this in my life," and "I never imagined that this could happen here."

Rare events are technically analyzed using discrete probability distributions. This should be done as a separate analysis from the risk analysis associated with those that can normally be expected. A common decision is whether or how to incorporate them into the risk evaluation and risk treatment components of the standard risk management framework.

Two general discrete probability distributions are the Binomial and the Poisson. The binomial is used to model whether a rare event will happen (a yes or no type approach). The Poisson is a limiting case of a Binomial distribution when the number of trials, n, gets large and p, the probability of success, is small. The Poisson distribution is often used to model the number of events (how many).

Monte Carlo analysis is often used to develop a project or management reserve, which is a planned amount of money or time to address normally unforeseeable situations. Risk evaluation and risk treatment determine the degree to which the probabilistic effects on equivalent financial costs and timing are included in the reserve fund, are transferred in the form of insurance to a third party, or simply taken at risk.

One major issue with the communication of rare events in the planning phase is that probabilistic analysis is not included in the risk analysis, and thus there is nothing to expressly communicate. The other major issue is that the communication aspects of the risk evaluation and risk treatment steps are simply not provided to an audience beyond the analyst or decision maker.

The foundational tool for the communication of the planning aspects related to rare events is a written risk register. Other communication approaches are discussed in the next chapter. Perhaps the most salient point here is that by their nature, rare events happen very infrequently and are usually beyond practical imagination. Considerations for rare events will be easily lost unless provided minimally in written form.

AFTERMATH OF A RARE EVENT

Rare events continue to unfold after the initial event. The chain of secondary events following the original disaster prolongs the impact of the disaster and introduces new risks. For example, structural damage to facilities incurred during an initial event leads to additional safety, environmental and public health risks until a system is brought back to its pre-event state.

All bets are off once the rare event happens. Those who believed that they were in control and had adequately planned understand that their assessment was inadequate. Those who were not in control of the planning efforts now see their fears being realized. And the degree of the realization will differ, either negatively or positively, from that which was originally expected.

Therefore, the public may accept a plan for addressing a rare event prior to its occurrence, but will reject it once they experience the actual event and cascading events in the aftermath. The need for communication is large during the recovery phase of a rare event because the deviation from expectations is large. It is especially important that communication be truthful, objective, and regular as the system which experiences a rare event bounces back to its pre-event state.

A more traditional approach to reliability, risk, and resiliency communication for a rare event is that the amount of noise is greatest in the planning phase. This is understandable given the different perspectives related to the low-likelihood, high-consequence events and the ability of those involved to accurately imagine the real impacts from descriptive explanations years before the actual event.

However, with the rise in the use of social media over the past decade, this traditional view is not necessarily true. The actual experience is that the public receives a continuous bombardment of information as soon as the event happens. Historically, official communication has balanced the need for timeliness and the need to be truthful and objective. Resiliency professionals must now have new communications strategies for reducing noise, and over-reaction, from continuous social media sources.

Risk and resiliency communication in the aftermath of a rare event must be dynamic and continuously updated. Although the potential for occurrence is lower than for other risks, pre-event communication planning is more important than for higher likelihood risks.

SUMMARY THOUGHTS ON RARE EVENTS

- Low-likelihood, high-consequence ("rare") events present special challenges to risk communications. Scientific and engineering rigor are essential for rare events, as with any risk assessment scenario. Managing the risks presented by rare events requires many of the same fundamental communication elements of any credible risk-based decision analysis. Given the diversity of stakeholders and unconventional aspects of rare events, a greater understanding and application of numerous other factors are needed, especially psychosocial and ethical factors.

- Resiliency is a key concept when considering and communicating risk related to rare events. Unlike regularly or normally occurring events, where reliability can be considered as a dominating concept to prevent occurrence, the nature of rare events makes them unexpectedly and disproportionately impactful. Making systems more resilient to low-likelihood, high-consequence events both addresses their actual survival and mitigates the psychological affects when they do occur.

CHAPTER 9

PLANS FOR COMMUNICATIONS

It is very important to have a plan for communicating reliability, risk, and resiliency to decision makers. Reliability, risk, and resiliency are complex in nature. They require analytical thinking. Relationships between input factors and outcomes usually have both overlapping and competing explanations. Processes involving the evaluation of reliability, risk, and resiliency are typically iterative in nature, and the evaluations take months and sometimes years to complete. The plan for communicating must be commensurate with the nature of the problem. It must be capable of efficiently and effectively getting the decision maker to make a decision—which, as we have discussed earlier in this book, is making an allocation of resources.

This chapter, and workshop module, belong at the end because it requires an understanding of all the material previously discussed. And yet I hate to place it here. My concern is that technical professionals due to their skill in logical thinking will conclude that a written "communication plan", often controlled by an organization's communication officer, is the necessary final objective for communicating effectively to decision makers. This is not the case.

The plan for communication begins on the day you begin working on a reliability, risk, or resiliency project. The planning effort is fluid. It is updated as more technical and non-technical aspects are known. Effective communication to decision makers happens throughout the project.

At the end of the technical effort, there will be a presentation that will likely take the form of a written report, verbal presentation, or both. For complex problems involving uncertainty, this presentation will be repeated several times and, in many cases, to people or groups that support the decision maker. There must be a *plan for communication* by the reliability, risk, and resiliency professional.

PRACTICALLY SPEAKING

I maintain that many technical professionals miss the obvious when it comes to developing their plan for communication. The obvious thing to me is to put yourself in the other person's shoes, or more simply, try to think like the receiver of the information. Another way to describe this is to have empathy for the decision maker. Many technical professionals have told me that this is difficult because theirs is the technical world and they have never been the chief decision maker or on a decision-making board. My response is, "You are making it too complicated."

All technical professionals have experienced someone trying to sell them something. When I do a workshop in conjunction with a professional conference, I usually ask my attendees what goes through their minds when they visit trade show booths.

First, we usually ask ourselves what affiliation the trade show salesperson has and if their affiliation is reputable. Second, we tend to ask ourselves what are the "angles" that the salesperson is trying to play to persuade us to come to their way of thinking. Third, and closely tied to the angles, is the inevitable thought of how this person benefits from convincing us. Fourth, we tend to explore their potential biases. That is, does the salesperson have an operational background? Maybe they do not have the perspective or training of an engineer. Maybe they are just pure sales. Our ultimate objective that follows from these thoughts is simply to decide whether we can believe them.

Practically speaking, you should let the decision makers know early in communication your credentials, your role in the decision-making process (including possibly how you benefit), your conclusions, and your underlying perspectives. This quickly gets the receiver past their natural first steps of evaluation, and removes another potential source that creates noise for your message. It also provides instant credibility. In most cases it establishes a sense of rapport between you (the sender) and your audience (the receiver) that will make communication throughout the rest of the presentation easier.

I took away three threats to credibility from an Edward Tufte workshop a few years ago. All three share the same underpinning. Tufte believes that once your credibility is gone, then you are worthless as a communicator to your audience.

Tufte's single biggest threat to credibility is "cherry picking." In other words, everyone has more data than they can possibly show or share. The issue is whether the data that is being presented is fair or is it selective to fit a position (i.e., cherry pick to fit the tastes and desires of the particular audience). The chapter on Ethics underscores this point that data should be presented in a fair and truthful manner.

Incompetence is Tufte's second biggest threat. For a presentation to the receiver, the presenter must be competent. The presenter must demonstrate knowledge of the subject matter and have some level of practical experience. Buzz words and jargon are no substitute for competence when addressing complex issues in the face of uncertainty. Establishing competence is best done early in the communication.

Tufte's third biggest threat to credibility is not understanding the context. I tend to explain this by reminding the reliability, risk, and resiliency communicator that their job is not to make the decision but to provide information to support the decision makers and the decision-making process. Tufte explains this as "rage to conclude." By this he means getting down to the essential reason you are there early. The core of the communication is to describe the problem, its relevance (who cares), and what actions/solutions are needed (what to do about it). Tufte also advices, analogous to the presentation of a technical paper at a conference, to start by reading the abstract.

Thus, the plan for communication is about establishing credibility, eliminating noise by quickly helping the receiver to accept your credibility, directing the decision maker's attention on the underlying information and analysis, and avoiding the threats to your credibility during your written or oral presentation. If you are credible and concise, then the decision makers will ask with great interest the many tiers of potential questions associated with your analysis that you hope they will ask. If you are not credible, it does not matter.

BEING CONCISE

I am a student of history. One of my favorite stories to share related to history and concise communications comes from Pauline Maier related to Thomas Jefferson, Benjamin Franklin, and the American Declaration of Independence. At the peak of drafting the Declaration, Jefferson had forgotten that he was a drafter, not an author, of what the Congress called the "declaration on independence." The declaration was not a novel, a poem, or a political essay by one author but a public document created by a body of people. Nevertheless, Jefferson sank into a state of depression as he saw his words, his draft, torn apart and rewritten by the group.

At that point, Jefferson recalled that Franklin attempted "to console him with the story of a young hatter, about to open his own shop, who proposed to have a fine signboard made with the words "John Thompson, Hatter, makes and sells hats for ready money" and the figure of a hat. First, however, he asked his friends for their advice. One proposed taking out "hatter" since it was redundant with "makes hats." Another recommended that "makes" be removed since customers wouldn't care exactly who made the hats. A third said "for ready money" was not necessary since it was not the local custom to sell on credit. "Sells hats," a fourth commented; did Thompson suppose people thought he meant to give them away? In the end, the sign said simply "John Thompson," with a picture of a hat, which probably served its function quite well" (Maier, 1997). Like Franklin's story, communicating reliability, risk, and resiliency should focus on make the message concise rather than the pride of authorship.

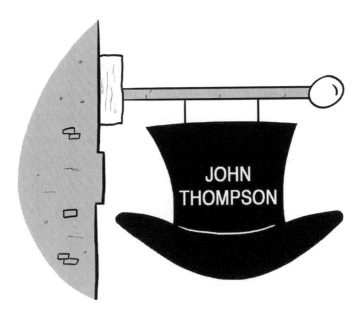

Keep it simple. Focus on the message and the needs of the receiver.

REACHING THE AUDIENCE

The most effective manner to reach the audience in terms of communicating reliability, risk, and resiliency is to get to the point (the results) early, let the data tell the story, and always be fair and truthful about the data and your analysis.

As discussed in an earlier chapter, most humans care about how decisions impact other people. Most humans also look for the comfort of established, proven processes that create consensus and help minimize unintended consequences or poor decisions. Addressing these aspects also provides secondary support for reaching the audience.

A key perspective that is frequently overlooked by technical professionals is that the organization of the presentation is important. Decision makers frequently do not sort through problems in the same way that technical professionals do. Technical professionals are trained in the scientific method and their analysis is based on a chronological order. In other words, we start with a hypothesis, develop a test plan, test the hypothesis, analyze the data, and conclude whether the hypothesis is valid. This logic builds from problem to conclusion over a period of time.

Decision makers often think in other sequences that are just as valid as chronological order. These include priority order, spatial arrangement, and problem-solution. The use of a risk-based prioritization model is an example where decision makers would prefer either a priority order or problem-solution sequence in a presentation to one based on chronological order. Many political and policy-making bodies prefer a spatial arrangement presentation format because it helps them to better understand and identify which of their constituents are most impacted by their decisions. Gaining the attention and keeping the attention of the audience is the responsibility of the sender, not the receiver, of the communication. Therefore, reliability, risk, and resiliency professionals must use sequences other than chronological order if they are to be effective communicators. Thus, your "how" is secondary to "what" you found as the result.

STANDARD COMMUNICATION PLANS

There are numerous standard communication plans that are available in texts and from the internet. In my workshops, I call these "communication plans for public relations and marketing" because their primary purpose is for advocacy and persuasion. Even though this strategy is different than the primary purpose of reliability, risk, and resiliency professionals, there are numerous good tips and pointers to be gained. I share a number of these standard communication plans with workshop attendees, including numerous ones I have authored for major environmental and infrastructure projects when filling the program communications role.

The Communications Action Plan for the $2 billion North Carolina International Port that I developed in conjunction with the State Ports Authority and two outside public relations firms is one I will share for one reason. The plan itself had the requisite standard components. We nicely rolled them up into three P's—Policy, Procedures, and People—for ease of use by the entire program team. The one part we did leave out under the People category was the profiling of the people we were to reach. It was an arguable point since most standard communication plans start with an early step of understanding your audience (recall the previous chapter that addressed personality profiles).

The big issue here is that most people on the receiving end of the communication do not want to be profiled at the individual level. One of my mentors, Roy Sowers, told me to keep the profile stuff out of print. Previously, he had to clean-up multiple situations where well-intended people put things in writing that was obtained by the media and the decision makers. It was never a pretty result.

One main difference in the "know your audience" aspect that is done in the marketing and public relations world, including politics, usually involves profiles for groups or classes of people. Although some people find even this to be offensive because of the associated stereotypes, it is generally more effective when framed and written in terms of mass marketing. What is universally offensive is that people find their profiles being distributed to a wide group of people to enact a strategy of persuasion or manipulation. This is intuitive when looking backwards, but I am amazed by the number of technical professionals I know that interpret "know your audience" to mean "profile your audience" and then to put it in writing and distribute it for all to see. Remember our focus as reliability, risk, and resiliency professionals is to communicate the data and our analysis in a fair and effective manner to the decision makers. We are not the decision makers. Our role is neither to persuade or manipulate the decision makers. Let the data and the analysis stand on its own.

Truth provides enough complexities and peril without throwing in an additional aspect of manipulation.

County Commissioner Frank Williams, also a public relations specialist, recently co-authored a paper at a utility management conference with me and my colleague, Adam Sharpe. It was targeted with helping public utility managers be more effective with their Boards. We boiled the standard communication plan down into just five parts for ease of discussion: communication goals and objectives; identification of the target audience; familiarity of the target audience with the subject matter (or technical analysis); frequency of communication; and key messages.

I share this format because it bridges the larger communication planning context, including the media and elected officials, with what I describe as the more focused communication that should be the approach of reliability, risk, and resiliency professionals. In this case, the primary focus is on the communication goals and objectives and the key messages. The secondary focus is the target audience, the target audience familiarity with the subject matter, and the frequency of communication.

I share examples from the public sector because these are, of course, public. One example is from a unit of local government which requires all technical communication that will enter a public forum to have a formal communication plan. It has 11 components and requires a minimum of eight pages to complete. Regardless of the good intentions and how much it conforms to a standard communications framework, this plan is too much work for front-line technical professionals.

LESSONS FROM CRISIS AND EMERGENCY COMMUNICATION PLANS

There are also numerous good examples in the public domain related to crisis and emergency communication plans. In my workshops, I reference *Crisis and Emergency Risk Communication (CERC)* from the Centers for Disease Control,

Federal Emergency Management Act (FEMA) *Effective Communications* manual and training, and the United States Environmental Protection Agency (USEPA) *Seven Cardinal Rules of Risk Communication.* All three are solid references that provide powerful insights for planning of communications. They are worthy of further discussion and study.

Their limitation is that they are largely focused on the aftermath of an event. Remember that in the aftermath of an unexpected event, System 1 thinking (fast, intuitive and emotional) is the rule of the day and human passion is largely at work. These references do have lesser relevance for pre-event planning and analysis, where System 2 thinking (slower, more deliberate, and more logical) dominates. However, as discussed earlier, the psychological effects of complexity and uncertainty often drive decision makers in the pre-event planning analysis mindset from the System 2 to the System 1 mindset.

From the CERC, I highlight the four ways people process information during a crisis:

- We simplify messages.

- We hold on to current beliefs.

- We look for additional information and opinions.

- We believe the first message.

The substitution of the term "complexity" or "uncertainty" for the term "crisis" makes the statement equally applicable to the human aspect of most reliability, risk, and resiliency decisions. In terms of this context, addressing the way people process information requires using simple messages, conveying messages from a credible source, using consistent messages, and releasing accurate messages as soon as possible.

The second topic from the CERC is the Mental States in a Crisis. There is uncertainty, hopelessness/helplessness, and denial. I contrast and compare these mental states in workshops with my own favorite, Kubler-Ross' five stages of grief (denial, anger, bargaining, depression, and acceptance). Reliability expert and good friend Tim Adams recently shared in one of the workshops his summary on the ways to respond to a risky situation being a mixture from different settings:

1. **Accept** (retain, engage, fight).

2. **Avoid** (run, flight).

3. **Hold** (freeze, but get more information for a next step).

4. **Mitigate** (change something to reduce the risk; countermeasure).

5. **Transfer** (share the risk).

Regardless of the paradigm, the essential point is that the plan for communicating reliability, risk, and resiliency to decision makers must anticipate these mental states and be prepared to address them. Understanding the decision context, getting to the point as quickly and concisely as possible, letting the data and analysis speak for itself, using an incremental or layered approach driven by audience questions, and respecting your audience are a handful of techniques to prepare for these mental states. Again, the burden is on the sender of the communication and not on the receiver.

FEMA's *Effective Communications* is a good overview of many of the topics we have already discussed. Due to its primary focus on emergency and post-emergency response, its major topics address sensory communications and communication mediums such as cell phones, radios, and social media.

USEPA provides the *Seven Cardinal Rules of Risk Communication*: accept and involve the public as a legitimate partner; listen to the audience; be honest, frank, and open; coordinate and collaborate with other credible sources; meet the needs of the media; speak clearly and with compassion; and plan carefully and evaluate performance. For planning communication for reliability, risk, and resiliency to decision makers, in workshops I emphasize three and discuss the underlying tips: listen to the audience (with seven supporting tips); be honest, frank, and open (with eight supporting tips); and speak clearly and with compassion (with 14 supporting tips).

RECOMMENDED COMMUNICATION FRAMEWORKS

Earlier, I referenced the six steps of decision quality from the Strategic Decision Group and Stanford University. In summary, these are:

- Frame the problem

- Creative, doable alternatives

- Meaningful, reliable information

- Clear values and tradeoffs

- Logically correct reasoning

- Commitment to action

I recommend this framework as a starting point for developing any technical approach related to reliability, risk, and resiliency. I similarly suggest using this framework as a final checklist to review any plan of communication, written or verbal, prior to delivery.

The quality decision process is only as good as its weakest link.

As a format or package for the plan of communication, I recommend the one my colleague Adam Sharpe and I have used successfully for many years for doing Business Case Evaluations (BCEs). Like standard communication plans, BCEs have many different formats depending on the organization and the author. Similar to standard communication plans, there are many texts and much information on the internet. A word of warning is that simply using any BCE format in planning for communication is no more acceptable than using any standard communication plan.

This recommended format includes six components:

- Conclusion and recommendation(s)

- Background

- Problem statement

- Alternatives

- Technical analysis

- Non-technical analysis

These six components should be covered in no more than two to four pages. Brief appendices should be attached to provide supporting information and to support additional questions. Typically, Appendix A provides support visuals and will include a geospatial map indicating where the problem is located, a process diagram, and one or two cross sectional depictions of the issue. Appendix B provides representative calculations and support graphics.

The goal is to provide a lean and concise written presentation of the data and the analysis. It should be able to be read by a decision maker in less than 10 minutes.

A common question is whether the conclusion and recommendation(s) should be in the beginning. The answer is yes. This is the message that the receiver (decision maker) needs to receive. It should be upfront and as separated as much as possible from any potential noise. Show respect for your audience. Answer their question first and trust them to be willing and able to finish reading a brief, concise report.

In example, Dianna Booher is among many communication experts that advocates for this approach. Booher describes getting to the message at the end of the presentations as the ascending structure. "Understanding a message written in this ascending format is an uphill climb" (Booher, 2001). Providing to the conclusions and recommendations at the beginning, or the descending approach, informs quickly and clearly. In addition to clarity, Booher cites reader control (the reader can read as much or as little as they wish) as a secondary reason from her preference of the descending format.

Another common question is whether recommendations should be provided. The answer to this is "it depends." First and foremost, the reliability, risk, and resiliency professional is not the decision maker. As an analyst, it is more than appropriate to provide conclusions and, as a matter of course, these will be expected since the analyst is responsible for turning data into the information that is needed to make a quality strategic decision. However, recommendations are situational. They are often a function of timing within the decision process. Discretion and judgement must be used as to whether to offer a recommendation versus a range of recommendations or a range of viable alternatives. I perform many probabilistic analyses so most of my of recommendations (or viable alternatives) are framed around probabilities, possibilities, and uncertainties.

VERBAL PRESENTATIONS

Three primary pointers that I provide are treat communication like it is a negotiation, listen to your audience, and know when it's time to leave.

By treating communication like a negotiation, I use two main pointers from negotiation training. The first is to project the problem onto a piece of paper or onto a screen via a projector. This takes away some of the personal issues that could exist when one individual presents something directly to another individual. It also symbolically focuses everyone's attention on the problem rather than the messenger. The second pointer related to negotiation training is to try to communicate with someone by sitting beside them rather than across the table. Over the past 30 years, I have been amazed by how many times I have seen staff or consultants on one side of the table and the decision maker(s) on the other. Symbolically it looks like two groups getting ready for battle, or two teams getting ready for a football game, rather than a cohesive team partnering to solve a problem.

Listening to your audience, and showing them respect, can take a number of different forms. The four forms I share are to be brief in the initial presentation, be ready with layers of backup information, truly listen to their issues and concerns (it is about them, not you), and pay careful attention to what is not said.

Knowing when it is time to leave is extremely important. The decision makers have a decision to make and usually want to have their debates without you there. Answer as many of their questions as possible directly and briefly. There is no need to try to look smart or overly impress the receivers, and certainly there is no place for lectures in communicating to decision makers. Be credible, say what you came to say, answer their questions, and leave. It is usually time to leave when they stop asking questions.

Finally, with respect to answering questions, I recommend a handful of key points. First, always repeat the question. It helps you verify that you are indeed about to answer the right question and it gives you a few seconds to think. Second, be brief by keeping your answer to only a few sentences. The decision maker will ask a follow-up question if you were not complete in your answer or if they want more detail. Third, when answering a controversial or contentious question, answer the person who asks the question directly but, after doing so, break away with your eye contact and pull in the other participants. Treating this like a negotiation, do not let the question and answer disintegrate into a one-on-one debate between you and the questioner. Fourth, stay on message by letting the data and the analysis stand on its own. There is no need to embellish it, and embellishing it will increase the probably of creating noise or misunderstanding. Finally, verify to the questioner if you indeed answered the question.

MORE FROM THE REAL WORLD

On the Friday prior to Super Bowl Sunday, a major water utility had to shut down its primary water plant that served its entire customer base. This included a major hospital and a major university. Its water system redundancy was in the form of a water line that interconnected with a neighboring community. Shortly after shutting down the water treatment plant, major failures began to occur within the backup water transmission system. Potable water was lost to the community for 24 hours and boiled water notices were not lifted until almost three days later.

Our team was called in to perform a root cause analysis on the Monday following the event and immediately after water services were returned to normal. Our focus was on the issues related to the water treatment plant. Another consultant was retained to do a similar analysis on the water transmission system that had failed. Both reports were due within four days and were to be released concurrently to the Board and the media on Friday at noon.

We saw the problem as two-fold. The first was obviously the technical analysis involving reliability, risk, and resiliency. A lot of smart people with engineering and operations competencies descended on the site and used several common analytical tools such as the 5 Why's, Causal Factor Charting, Failure Modes and Effects Analysis (FMEA), models, and statistical data analysis.

Equally important, we saw the second part of the problem as being able to communicate the technical findings effectively to decision makers (and in this case also the media). Our report had to be brief and concise. And it was going to the administration, the Board, and the media simultaneously. No second chances, no re-writes, and no outside reviews.

There was no competition between the two independent firms who performed the technical analysis on different parts of the system. Both reports independently made the noon Friday deadline. Both reports were solid from a technical perspective. And both went directly into the public domain.

Our report was highly complimented by senior management and the board of directors for its conciseness and ease of understanding. It was directly duplicated in large part by the printed media in doing their articles. The media often is a little wrong in these type of complex technical matters, but this time they got it right. When I returned a call to one of the primary reporters the next day, she said the report provided her exactly what she needed.

The communication aspect was missing in the other consultant's report. In fact, their report was so technical that a supplemental "layman's summary" had to be released later in the day for the board and the media. That layman's summary missed the media deadline on that Friday afternoon, and was even referenced as pending in the major news article on issue. They got the technical part right, but the equally important communication piece was missing.

This simple story brings us back to where we started. The genesis of this book dates back about four years. In searching for interesting and relevant risk and reliability topics for papers and presentations, I asked numerous risk and reliability professionals what was the hardest part of their job. I got an almost universal reply that help was needed "to get senior management to understand what I do." When I questioned this answer, the reply again was almost universally "I have the tools and knowledge to do the technical part; I just need to get them to understand." This book and my workshops have been aimed at filling this gap in the profession. I hope this book has been helpful for you.

SUMMARY THOUGHTS ON PLANS FOR COMMUNICATION

- For reliability and risk, it is less about the colors, layout, and sequencing. It is mostly about getting to the point and letting the data be the primary focus for communicating about complex decisions under uncertainty. Colors, layout, sequencing and similar considerations are secondary to the data and related key messages.

- First communicate the recommendations and conclusions and be brief in the overall presentation. Be comfortable letting the decision maker ask questions if they need further explanation. It is about them (the receiver), not you (the sender).

- Standard plans and approaches for public relations and marketing are not aligned with some of the most important features of communicating reliability, risk, and resiliency. The Six Steps of Decision Quality or most Business Case Evaluation (BCE) formats provide good default templates for using text.

- The most important aspect of the question and answer period is listening before speaking. Be sure to confirm that you have answered the decision maker's question and concern.

THE END

NOTES

CHAPTER 1-RELIABILITY, RISK, AND RESILIENCY

Reliability Definitions: The cited definition of reliability is from MIL-STD-721C (1981): Military Standard: Definitions of Terms for Reliability and Maintainability. O'Connor and Kleyner (2012) define reliability in a very similar manner to MIL-STD-721C, but add "...function *without failure*..." Bazovsky (1961) is one of the earliest reliability texts in the modern era and defines reliability in a very similar manner. Linn's article, *History of Reliability Engineering*, provides additional insights into the definition and use of the term reliability (http://www.asqrd.org).

Risk Definitions: The update to ISO 31000:2009 is expected to be released in 2018. ISO Guide 73:2009 also provides the definitions of generic terms related to risk management, and in the future all ISO risk standards will reference this document as a common standard. Bernstein's *Against the Gods: The Remarkable Story of Risk* (1996) remains the standard text for understanding the history and usage of the term risk. Knight (1921) provides the basis for the modern definition of risk and uncertainty.

Risk versus Uncertainty: See Volz and Gigerenzer (2012) for an interesting and one of the most concise discussions of the topic. The list of traditional works and authors provided in the Notes on Chapter 2 under the heading of Decision Making provides more references on the topic, including Knight, L.J. Savage, S.L. Savage, Arrow, Kahneman, Tversky, and Slovic.

Resiliency Definitions: The definition of resiliency is sourced to the Merriam-Webster Dictionary and the American Society of Civil Engineers. Fiksel, Goodman, & Hecht (2014) are referenced as an expansion of the definition of resiliency through adaptive management. Resiliency as a term has grown to be almost interchangeable with "sustainability" in the environmental community.

Operational Resiliency: David Wilbur describes operational resilience as the ability for errors to be absorbed into a man-made system without compromising its mission. The foundation for this perspective is based on several converging fundamentals: the loss of reliability in complex systems is usually the result of human error; humans can "learn as they go" and adapt their processes to absorb more human error; and there is no such thing as eliminating all human errors. In concept, resiliency improvement (or growth) is analogous to the traditional concept of reliability growth. However, the point to be made here is that the definitions of reliability and risk have been well established for many years while the definition of resiliency, either related to natural or man-made systems, is still evolving. See Wilbur (2017).

CHAPTER 2-OTHER CONCEPTS THAT CONFUSE

The Chance of Rain: Gigerenzer's *Risk Savvy* (2014) and Joslyn, Nadav-Greenberg, and Nichols' article *Probability of Precipitation* (2009) are the primary sources of the understanding of the probability of rainfall. NOAA and NASA also are good sources for understanding weather forecasts.

Understanding Probability: Gigerenzer (2002, 2014), Gigerenzer and Edwards (2003), and Kahneman, Tversky, and Slovic (2009) are good sources for better understanding our lack of instinct for understanding probability. See also Spetzler and von Holstein (1972). Leonard "Jimmie" Savage (1954) provides interesting insights.

Decision Making: Kahneman (2012) provides the best overall summary of decision making to that point in time. System 1 and System 2 are attributed to Stanovich and West (2000), although Kahneman also credits Seymour Epstein, Jonathan Evans, and Sloman (1996) as pioneers in the field. Traditional works on decision making such as Arrow (1951), Meehl (1954), Raiffa (1968), and Hammond, Keeney, and Raiffa (1999) are worth a read. Taleb (2004, 2010) also provides good references to many historical works on decision making. More modern classics on decision making include Kahneman, Slovic, and Tversky (1982), Dawes (1979), Keeney (1992), Kleindorfer, Kuenruether, and Schoemaker (1993), Russo and Schoemaker (2002), and Slovic (2000). Edwards, Miles, and Winterfeldt, editors (2007) provide the best reference of essays and papers on the subject. The six-step decision process referenced here is from Stanford University, Stanford University Center for Professional Development, and Strategic Decisions Group International LLC (2012), and is also referenced by Schoemaker (2009).

Complexities of Decision Making: A number of academic models exist. For example, Parnell describes a four-dimensional model that includes: number of decision makers; level of stakeholder interaction; value and time preference; and uncertainty and risk preference (Parnell, Gregory S. (2009). Decision Analysis in One Chart. *Decision Line*, May 2009).

Definition of a Decision: From Stanford University, Stanford University Center for Professional Development, and Strategic Decisions Group International LLC (2012).

Difficulty Making Decisions, Part 1: Summarizing this into two issues is based on personal observation. See sources referenced under Decision Making for a broader range of opinion on why people have a hard time making strategic decisions. Stock A and Stock B reference can be found in Kahneman, Slovic, and Tversky (1982) and Kahneman (2011).

Difficulties Making Decisions, Part 2:   One interesting perspective related to senior and middle managers comes from Nutt, who identified four methods used to make strategic decisions. Data-based and collaboration-based methods were effective more than 90% and 80% of the time, respectively, while edict-based and persuasion-based methods were effective less than 50% of the time. However, the two most effective methods were used by top managers 21% of the time. Nutt concludes that the problem is that the best tactics take time to implement and many managers are looking for quick fixes (Nutt, P.C. (1998). Evaluating Complex Strategic Choices. *Management Science*, 44 (8), 1148-1166).

Risk Management Frameworks: See ISO 31000:2009 (2009) and COSO (2016).

Managing Risks: Tim Adams describes five ways to respond to risk, and his method incorporates three basic choices being "fight", "flight", or "freeze". The four according to De Loach are "avoid", "reduce", "transfer", and "accept" or retain (De Loach, J. W. (2000). Enterprise-wide Risk Management: Strategies for linking risk and opportunity. London: *Financial Times*/Prentice Hall). De Loach also added "exploit" as an additional option. Tomlin added "ignore" (Tomlin, B. (2006). On the Value of Mitigation and Contingency Strategies for Managing Supply Chain Disruption Risks. *Management Science*, 52 (5)).

CHAPTER 3 - GRAPHICAL EXCELLENCE

Graphical Excellence: Tufte's primary works are referenced here, including his four major books (1990, 1997, 2001, 2007). His website is www.edwardtufte.com and is provided in the workshops.

Lying with Data: Huff (1954) and Monmonier (1991) are primary examples for general reference.

Core Graphics: In addition to references provided by Tufte, see also references such as Juran (1998) and Pyzdek and Keller (2013) related to Total Quality Management. Tim Adams points out that different sources describe seven, but the seventh is described differently depending on the source.

Checklists: See Gawande (2009) for a good discussion of the use of checklists.

Risk Professionals, Risk Matrices, and Heat Maps: See Solomon, Vallero, and Benson (2017) for a discussion of the scales-statistic controversy. See Cox (2008) and Hubbard (2009) for a focused discussion on the challenges of risk matrices.

Visuals Used in Training and Learning: See Clark (2010). Clark is an excellent source in the areas of training and learning using different architectures and media.

Colors: There are many sources that describe the use of colors and their use in marketing, branding, and business. One source is Lewis (2014). See also Clark and Lyons (Colvin, Ruth and Lyons, Chopeta (2003). More Than Just Eye Candy: Graphics for e-Learning: Part 1 and Part 2, *Learning Solutions Magazine*. Retrieved on-line www.learningsolutionsmag.com/).

Graphics rather than Words: I was reminded recently at a workshop that dyslexia is another good reason to keep written messages simple and to use alternative means to written communication, including graphics, as a means for effectively conveying a message. Dyslexia is a variable learning disability involving difficulties in acquiring and processing language that is typically manifested by a lack of proficiency in reading, spelling, and writing. Many people suffer from this disability. Like color blindness, this reminds us to keep our messages and our graphics as simple and concise as possible if we are to ensure that they are effectively received.

CHAPTER 4-ETHICS

Type of Ethics: See Vallero (2007a, 2007b) and McCloskey (2014).

Minimalist Ethics: See Vallero (2007a, page 36 of Biomedical Ethics for Engineers). Minimalist ethics following the "malpractice model" are, in part, comprised of both Mill's harm principle and Hobbes' social contract. It also draws several parallels with Kohlberg (i.e. the licensing of engineers is preconditional or conditional, at best—see Fig. 2.8 in Biomedical Ethics).

Professional Ethics: Professional Codes of Ethics are readily available on websites. See American Society of Civil Engineers (2016) as an example.

Codes of Ethics: Dan Vallero reminds me that it is interesting that even though the engineering code of ethics is written mainly as proscriptive and deontological, the first canon is really about virtue, i.e. hold paramount public health, safety, and welfare. The code of ethics, like most humans, has elements of all three types of ethics. Be virtuous, the outcomes must be of positive consequence to the public, and carry out your work so as not to violate any of the duties of our profession.

CHAPTER 5-PRACTICAL TOOLS

Palisade DecisionTools Suite: Palisade's DecisionTools Suite is an excellent set of technical tools that have been well developed for effective communication. Palisade has remained committed to Microsoft Excel as a universal platform, so tools and supporting graphics are easy to share. Website is www.palisade.com.

Trouble Explaining Histograms: In addition to the effectiveness of comparative analysis that was referenced in this chapter, Tim Adams reports he is effective by explaining to the decision maker that any data set has at least two, and no less than two, measures. These are central tendency (e.g., mean, median, mode) and variation or dispersion (e.g., standard deviation, range). With that in mind, a histogram's primary purpose is to depict variation. I agree with Tim on his key points but have had a very different experience in trying to describe effectively in a short period of time concepts such as central tendency and variation.

Examples: This chapter is focused on real world examples which are shared in the workshops. See also Solomon and Sharpe (2013, 2015) and Solomon (2014).

CHAPTER 6-NOISE

Communication Defined: The definition and elements of communication are based on discussions with one of my primary mentors, Colonel Roy G. Sowers, and his references from the US Army. See US Department of the Army (1984).

Noise: The discussion of noise in communications being analogous to signal-to-noise ratio is based on discussions and collaborations with Daniel Vallero. See Solomon and Vallero (2016). The three forms of communication are after Vallero (2007), Myers and Kaposi (2004), and Green (1989).

Definition of Noise: The provided definition in the chapter is one common definition from electrical and communications engineering. I agree with Tim Adams that a practical definition for technical professionals is that noise is anything that degrades or distracts from providing clear communication. It is not sufficient to say the receiver feels good about what was received and/or is willing to commit to action.

Discussions of Noise: Describing noise in four components is the most effective way that I have found to cover the topic in workshops.

These are forms, forums (workshops, public speaking, media events, and elected officials), image (visual, written vocal), and patterns.

Visual, Written, and Vocal Image: See Booher (2009) for 101 of the most common grammatical errors. Clark (2010) provides more insights into image as part of her three training components: delivery modes (devices); communication modes (visual, text, audio); and training methods (techniques).

Forms of Communication: Numerous text are available related to public speaking, communicating with the media, and communicating with elected officials. These are not listed in detail here. Blackwell (1998) is one source for solid reference material in this area.

Patterns of Communication: See Leavitt (1951).

CHAPTER 7-THE AUDIENCE

Personality Styles: Also known as personality profiles. See Jung (1923), and Marston (1928) for the source documents.

Handwriting Analysis: Seifer (2009) does a reasonable job capturing the history and range of handwriting analysis. McNichol (1994) has a good, practical book. I prefer the passion and down-to-earth approach of Baggett (1998) and good resources can be found at his website and the Handwriting University.

Neuro-Linguistic Programming (NLP): See Bandler and Grinder (1979). Knight (1995) has a good overview. The internet is filled with many debates on the validity of the methodology. See Sturt, et al (2012) for one of the few recent academic-type reviews.

Individual Biases: In addition to Kahneman, Tversky, and Slovic, see also Gilovich et al, editors (2002) for an outstanding collection of the leading papers and essays on heuristics and biases.

Group Behavior: Fried (1976), Talbot (2010), Cyert and March (1992), Simon (1997), Likert (1967), Argyris (1957), and

Odiorne (1987). See Natemeyer and Hersey (2011) for a collection of classic essays on organizational and group behavior.

CHAPTER 8-RESILIENCY AND RARE EVENTS

The Nature of Rare Events: My discussions and collaboration with Dan Vallero helped to consolidate much of my thinking related to rare events. See Vallero (2007). Mitroff (2000) and his website (https://mitroff.net/) are good sources of information. Crisis management is often in the realm of security and emergency response professionals while the planning and treatment of rare events falls in the realm of risk and reliability professionals. Therefore, the terminology may be different but most of the core considerations are the same. Drawing analogy of rare events to physical events associated with crisis and the psychological reactions in the aftermath of crisis has been especially effective with audiences.

Stress-Strain Relationships, Unreliability, and Risk: I am very appreciative of discussions with Tim Adams that more formally addressed my instinctive thinking on this topic, especially given the fact that I used it informally for several years without much technical backing. More information can be found by searching on the website's name "KSC Reliability" or using https://kscddms.ksc. nasa. gov/ Reliability/

Prospect Theory: See Kahneman and Tversky (1979) and Kahneman (2011)

Perception of Risk: See Slovic (1999), Slovic (2000), Slovic and Weber (2002), Tversky (2004), and Kahneman (2011).

Aftermath of a Rare Event: See Solomon and Vallero (2016).

CHAPTER 9-PLANS FOR COMMUNICATIONS

Practically Speaking: The majority of this topic is from much severe personal experience. The works and workshops of Edward Tufte also summarize this topic well.

Teaching and Presentations: Tim Adams pointed out to me that engineers are not simply biased to a chronological view because of bias or alignment with the scientific method. Traditional educational approaches are based on the underlying tenant that we learn in a bite-size, cumulative manner. Our default toward chronological organization of presentation is at least partially rooted in the way we were educated from a very early age.

Presentation Organization: A number of good sources are available in the market. Some of my older default references include Arredono (1991), Leech (1993), Mandel (1987), and Monkhouse (1981).

Crisis and Emergency Response Communication: See CDC (2017), FEMA (2017), and USEPA (1998) for good foundational sources of information.

Business Case Evaluations: There are many formats that are used in the market for Business Case Evaluations (BCE) and Business Case Analysis (BCA). My colleague Adam Sharpe and I have developed a format which works well in our practice related to risk, reliability, asset management, and decision analysis. The core aspects are provided in the text and workshop. However, our 8-hour training module has not yet been converted into a publicly available document.

Components of R&M Analysis: Tim Adams uses a six-component approach to reliability and maintainability (R&M) analysis: Scope/boundaries; data sources; scenarios; method; findings and conclusions; and recommendations.

Booher and the MADE Format: Booher (2001) created the MADE format. M is for Message, which should be the stated bottom-line message that interest the audience. A is for Actions, which should be taken by you or the audience. D is for Details. E is for evidence, which includes any attachments or handouts that you will be leaving with the audience or forwarding later. MADE is most applicable to emails, but it also a good fundamental approach for technical reports and presentations.

Fielding Live Questions: There are numerous sources available in the market for more information. The ones provided in this text and the workshops are the top ones that I found most relevant over the years. My list is by no means exhaustive.

Communicating Information Effectively: Tim Adams uses what he describes as the R2D2 (Read, Review, Discuss, and Decide) method for sharing information.   First information is sent to participants in advance of a face-to-face meeting. The information is reviewed at the start of the meeting. If during review it becomes clear that a participant has not read the information, discussion from that participant stops. The end goal is to reach a decision on the provided information following discussion. The final decision step should not be performed if the previous three are not performed in sequence. Tim uses this technique primarily as a training method for mastery. It also is effective in making business meetings more effective and communicating information more effectively (reducing noise).

GLOSSARY
QUICK, STANDARD DEFINITIONS

Understanding derived from concise definitions is critical for quality decision making. The following are some quick, standard definitions that have been cited in this book. The majority are sourced to US Military Standards, such as MIL-STD-721C (Reliability Terminology) or to ISO Standards, such as ISO 31000 (Risk), ISO 55000 (Asset Management), or ISO 15288 (Systems Engineering). Others are sourced to professional society publications (such as the American Society for Quality or the American Society of Civil Engineers or the Society of Automotive Engineers) or US government agencies (such as the United States Environmental Protection Agency). In a few cases, the exceptions, definitions are sourced to specific thought leaders.

Technical professional will continue to debate the fine points related to these definitions. This book is not intended to resolve these debates. For communications by reliability, risk, and resiliency professionals to decision makers, it is more important that a concise, single definition be provided, regardless of it source, to achieve a common understanding among communication receivers.

Availability: The probability that an item is capable of performing its function under stated condition at a random time.

Communication: The exchange of information or ideas from one person to another.

Decision: An irrevocable (or irreversible) choice among alternative ways to allocate resources.

Dependability: The probability that an item will meet its intended function during its mission.

Failure: The event, or inoperable state, in which any item fails to do what it is intended (specified) to do.

Function: What the owner or user of a physical asset or system wants it to do, normally expressed by a noun, verb, and performance level.

Maintainability: The probability that an item can be retained or restored to its specified condition when maintenance is performed by personnel having the specific skill, and using prescribed procedures and resources.

Noise: Anything that degrades or distracts from providing clear communication to the receiver.

Probability: The mathematical measure of the chance of occurrence expressed as a number between 0 and 1, where 0 is impossibility and 1 is absolute certainty.

Reliability: The probability that an item will perform its intended function for a specified interval under stated conditions.

Resiliency: The ability to return to the original form or state after being stressed

Risk: The effect of uncertainty on objectives.

System: A group of interdependent components, processes and people that together perform a common purpose or mission.

BIBLIOGRAPHY

American Society of Civil Engineers (2016). *Code of Ethics*. ASCE. Retrieved from http://www.asce.org/code-of-ethics/

American Society for Quality (2017). *The 7 Basic Quality Tools for Process Improvement*. ASQ. Retrieved from http://asq.org/learn-about-quality/seven-basic-quality-tools/

Argyris, Chris (1957). *Personality and Organization: The Conflict Between System and the Individual*. New York: Harper & Row, Publishers.

Arredono, Lani (1991). *How to Present Like a Pro*. New York: McGraw-Hill Inc.

Arrow, Kenneth Joseph (1951). *Social Choices and Individual Values*. New York: John Wiley & Sons, Inc.

Baggett, Bart (1998) *The Secrets to Making Love Happen: Mastering Your Relationships Using Handwriting Analysis and NLP*. New York: Sterling Publishers.

Bandler, R., Grinder, J., & Andreas, S. (1979). *Frogs into Princes: Neuro Linguistic Programming*. Moab, Utah: Real People Press.

Bazovsky, Igor (1961). *Reliability Theory and Practice*. Mineola, New York: Dover Publications.

Bernstein, Peter (1996). *Against the Gods: The Remarkable Story of Risk*. New York: John Wiley & Sons, Inc.

Bier, Vicki M., and Cox, L.A. (2007). Probabilistic Risk Analysis for Engineered Systems, *Advances in Decision Analysis: From Foundations to Applications*, Edwards, Miles Jr., & von Winterfeldt, editors.

Cambridge: Cambridge University Press.

Blackwell, Joel (1998). *Personal Political Power: How Ordinary People Get What They Want from Government.* Essex, Connecticut: Issue Management Co. LLC.

Booher, Dianna (2001). *E-writing: 21st-Century Tools for Effective Communications.* New York: Pocket Books.

Booher, Dianna (2009). *Booher's Rules of Business Grammar: 101 Fast and Easy Ways to Correct the Most Common Errors.* New York: McGraw-Hill.

Centers for Disease Control (2017). *Crisis and Emergency Risk Communication (CERC).* Retrieved from https://emergency.cdc.gov/cerc/

Clark, Ruth Colvin (2010). *Evidence-Based Training Methods: A Guide to Training Professionals.* Alexandria, VA: ASTD Press.

Council of Supporting Organizations (COSO) of the Treadway Commission (2016). *Enterprise Risk Management — Aligning Risk with Strategy and Performance.* Retrieved from https://www.coso.org/

Cox, L.A. (2008). What's Wrong with Risk Matrices? *Risk Analysis,* Vol. 28, No. 2.

Cyert, Richard and March, James (1992). *A Behavioral Theory of the Firm, 2nd Edition.* Oxford, England: Blackwell Publishes Ltd.

Dawes, Robyn (1979). The Robust Beauty of Improper Linear Models in Decision Making. *American Psychologist,* July 1979, Vol. 34, No. 7.

Edwards, Ward, Miles Jr., Ralph F., and von Winterfeldt, editors (2007). *Advances in Decision Analysis: From Foundations to Applications.* Cambridge: Cambridge University Press.

Federal Emergency Management Act (2017), IS-242.B: *Effective Communications.* Retrieved from https://training.fema.gov/

Fiksel, J., Goodman, I., and Hecht, A. (2014). Resilience: Navigating toward a Sustainable Future. *Solutions for a Sustainable and Desirable Future,* 5(5), 38-47. Retrieved from http://www.thesolutionsjournal.com/ node/237208

Fried, Robert C. (1976). *Performance in American Bureaucracy.* Boston: Little, Brown and Company.

Gawande, Atul (2009). *The Checklist Manifesto.* New York: Metropolitan Books.

Gigerenzer, Gerd (2002). *Calculated Risks: How to Know When Numbers Deceive You*. New York: Simon & Schuster.

Gigerenzer, Gerd (2014). *Risk Savvy: How to Make Good Decisions*. New York: Penguin Group.

Gigerenzer, Gerd and Hoffrage, Ulrich (1995). How to Improve Bayesian Reasoning without Instruction: Frequency Formats. *Psychological Review*, VoTl02, No. 4, p. 684-704.

Gigerenzer, Gerd, and Edwards, Adrian (2003). Simple Tools for Understanding Risks: From Innumeracy to Insight, *British Medical Journal*, 327.

Gilovich, Thomas, Griffin, Dale W., and Kahneman, Daniel, editors (2002). *Heuristics and Biases: The Psychology of Intuitive Judgement*. New York: Cambridge University Press.

Green, T. R. (1989). Cognitive Dimensions of Notations. *People and Computers V*, 443-460.

Hammond, John S, Keeney, Ralph L., and Raiffa, Howard (1999). *Smart Choices: A Practical Guide to Making Better Life Decisions*. New York: Broadway Books.

Hubbard, Douglas W. (2009). *The Failure of Risk Management: Why It's Broken and How to Fix It*. New York: John Wiley & Sons, Inc.

Huff, Darrell (1954). *How to Lie with Statistics*. New York: W.W. Norton & Company, Inc.

International Organization for Standardization (ISO) (2009). *ISO 31000-2009: Risk management — Principles and Guidelines*. Geneva. www.iso.org.

Joslyn, Susan, Nadav-Greenberg, Limor and Nichols, Rebecca M. (2009). Probability of Precipitation: Assessment and Enhancement of End-User Understanding, *American Meteorological Society*.

Jung, Carl G (1910). The Association Method. *American Journal of Psychology*, 31, 219-269.

Jung, Carl G (1923). *Psychological Types*. Princeton, New Jersey: Princeton University 191 edition.

Juran, Joseph (1998). *Juran's Quality Handbook*, 5th edition, New York: McGraw-Hill.

Kahneman, Daniel (2011). *Thinking, Fast and Slow*. New York: Farrar, Straus and Giroux.

Kahneman, Daniel and Tversky, Amos (1979).   Prospect Theory: An Analysis of Decision under Risk", *Econometrica* (pre-1986); Mar 1979; 47, 2; ABI/INFORM Global.

Kahneman, Daniel, and Lovallo, Dan (1993). Timid Choice and Bold Forecasts: A Cognitive Perspective on Risk Taking. *Management Science*, Vol 39, No 1.

Kahneman, Daniel, Slovic, Paul, and Tversky, Amos (1982). *Judgment Under Uncertainty: Heuristics and Biases.* New York: Cambridge University Press.

Keeney, Ralph L. (1992). *Value Focused Thinking: A Path to Creative Decisionmaking.* Cambridge: Harvard University Press.

Kleindorfer, Paul R, Kunreuther, Howard C., and Schoemaker, Paul J.H. (1993). *Decision Sciences: an integrative perspective.*   New York: Cambridge University Press.

Knight, Frank (1921). *Risk, Uncertainty, and Profit.* Boston: Houghton Mifflin Company.

Knight, Sue (1995). *NLP at Work: The difference that makes a difference in business.* London: Nicholas Brealey Publishing.

Leavitt, Harold J. (1951). "Some Effects of Certain Communication Patterns on Group Performance", *The Journal of Abnormal and Social Psychology*, Vol 46(1)

Leech, Thomas (1993). *How to Prepare, Stage, and Deliver Winning Presentations.* New York: American Management Association.

Lewis, Richard (2014). *Color Psychology: Profit from the Psychology of Color: Discover the Meaning and Effect of Colors.* Riana Publishing.

Likert, Rensis (1967). *The Human Organization: Its Management and Value.* New York: McGraw-Hill, Inc.

Maier, Pauline (1997). *American Scripture: Making the Declaration of Independence.* New York: Alfred A. Knopf, Inc.

Mandel, Steve (1987). *Effective Presentation Skills.* Los Altos, California. Crisp Publications.

Marston, W. M. (1928). *Emotions of Normal People.* London: K. Paul, Trench, Trubner & Co. Ltd.

McCloskey, H.J. (2014). *Meta-Ethics and Normative Ethics.* New York: Springer.

McNichol, Andrea (1994). *Handwriting Analysis: Putting It to Work for You,*

1st Edition. New York: McGraw-Hill Education.

Meehl, Paul E. (1954). *Clinical versus Statistical Prediction: A Theoretical Analysis and a Review of the Evidence*. University of Minnesota.

Merriam-Webster's Collegiate Dictionary (10th ed.) (1993). Springfield, MA: Merriam-Webster.

Mitroff, Ian I. (2000). *Managing Crisis Before They Happen: What Every Executive and Manager Needs to Know About Crisis Management*. New York: AMACOM.

Monkhouse, Bob (1981). *Just Say a Few Words: The Complete Speaker's Handbook*. New York: Dorset Press.

Monmonier, Mark (1991). *How to Lie with Maps*. Chicago: University of Chicago Press.

Montgomery, Doulas C., Peck, Elizabeth A., and Vining, G. Geoffrey (2012). *Introduction to Linear Regression Analysis, fifth edition*. New York: John Wiley & Sons, Inc.

Myers, M., & Kaposi, A. (2004). *A first systems book: Technology and management*. Imperial College Press.

Natemeyer, Walter E. and Hersey, Paul (2011). *Classics of Organizational Behavior, fourth edition*. Long Grove, IL: Waveland Press, Inc.

Odiorne, George S. (1987). *The Human Side of Management*. Lexington, Massachusetts: D.C. Heath and Company.

Palisade Corporation, website www.palisade.com, 2013, 2016, 2017.

Patrick O'Connor and Andre Kleyner (2012). *Practical Reliability, Fifth Edition*, John Wiley & Sons.

Project Management Institute (2004). *Guide to the Project Management Institute Body of Knowledge*. Newtown Square, Pa: Project Management Institute.

Pyzdek, Thomas, and Keller, Paul (2013). *The Handbook for Quality Management. A Complete Guide to Operational Excellence*, second edition. New York: McGraw-Hill Education.

Raiffa, Howard (1970). *Decision Analysis, Introductory Lectures of Choices Under Uncertainty*. Reading, Massachusetts: Addison-Wesley Publishing Company.

Russo, J Edward and Schoemaker, Paul J.H. (2002). *Winning Decisions*. New York: Currency Doubleday.

Savage, Leonard J. (1954). *The Foundations of Statistics*, New York: John

Wiley & Sons.

Savage, Sam L. (2009). *The Flaw of Averages: Why We Underestimate Risk in the Face of Uncertainty.* New York: John Wiley & Sons, Inc.

Schoemaker, Paul J.H. (2009). Implications of Decision Psychology for Classical Notions of Rationality, in *The Irrational Economist: Future Directions in Behavioral Economics and Risk Management*, E. Michel-Kerjan and P. Slovic (eds.) Public Affairs Press.

Seifer, Marc (2009). *The Definitive Book of Handwriting Analysis.* Pompton Plains, NJ: The Career Press Inc.

Simon, Herbert A. (1997). *Administrative Behavior: A Study of Decision-making Processes in Administrative Organizations, fourth edition.* New York: The Free Press.

Sloman, Steven (1996). The Empirical Case for Two Systems of Reasoning. *Psychological Bulletin by the American Psychological Association, Inc.*, Vol. 119. No. 1.

Slovic, Paul (1999). Trust, Emotion, Sex, Politics, and Science: Surveying the Risk-Assessment Battlefield. *Risk Analysis,* 19(4), 689-701.

Slovic, Paul, and Weber, E. (2002). *Perspective of risk posed by extreme events.* Paper presented at the conference 'Risk Management Strategies in an Uncertain World'.

Slovic, Paul (2000). *The Perception of Risk.* New York: Taylor & Francis.

Solomon, J.D., Vallero, Daniel A., Benson, Kathryn (2017). *Evaluating Risk: A Revisit of the Scales, Measurement Theory, and Statistical Analysis Controversy.* Proceedings from the 2017 Reliability and Maintenance Symposium (RAMS 2017).

Solomon, JD and Vallero, Dan (2016). Communicating Risk and Resiliency: Special Considerations for Rare Events. *The CIP Report.* The Center for Infrastructure Protection and Homeland Security, George Mason University, https://cip.gmu.edu.

Solomon, JD, and Sharpe, Adam (2013). *Regional Planning Using @RISK, PrecisionTree and Evolver, Part 2: Communicating to Decision Makers.* Proceedings from The Risk Conference. Las Vegas, NV.

Solomon, JD (2014). *Using the Palisade Suite for Risk and Reliability Optimization of Infrastructure Systems.* Proceedings from The Risk Conference, New Orleans, LA.

Solomon, JD, Sharpe, Adam, and Seachrist, Steven (2014). *The Leading*

Edge: Gwinnett County DWR'S Approach to Infrastructure Renewal and Replacement. Proceedings from the AWWA ACE Conference, 2014.

Solomon, JD, and Sharpe, Adam (2015). *Tailoring Analytical and Communication Approaches for High Stakes Capital Infrastructure Projects.* Proceedings from The Risk Conference. Chicago, IL.

Spetzler, Carl S., and Stael von Holstein, C.S. (1975). Probability Encoding in Decision Analysis. *Management Science*, Volume 22(3).

Spetzler, Carl S. (2007). Building Decision Competency in Organizations, *Advances in Decision Analysis: From Foundations to Applications,* Edwards, Miles Jr., & von Winterfeldt, editors. Cambridge: Cambridge University Press.

Stanford University, Stanford University Center for Professional Development, and Strategic Decisions Group International LLC (2012). Stanford Strategic Decision and Risk Management Certification Program.

Stanovich, Keith E. and West, Richard F (2000). Individual Differences in Reasoning: Implications for the Rationality Debate. *Behavioral and Brain Sciences*, 23.

Sturt, Jackie, Saima, Ali, Robertson, Wendy, Metcalfe, David, Grove, Amy, Claire Bourne, Claire, and Chris Bridle, Chris (2012). Neurolinguistic Programming: A Systematic Review of the Effects on Health Outcomes, *British Journal of General Practice*, 2012 Nov; 62(604): e757–e764. Published online 2012 Oct 29.

Talbot, Colin (2010). *Theories of Performance: Organizational and Service Improvement in the Public Domain.* Oxford: Oxford University Press.

Taleb, Nassim Nicholas (2004). *Fooled by Randomness.* New York: Random House.

Taleb, Nassim Nicholas (2010). *The Black Swain: The Impact of the Highly Improbable.* New York: Random House.

Tufte, Edward (1990). *Envisioning Information.* Cheshire, Connecticut: Graphics Press.

Tufte, Edward (1997) *Visual Explanations.* Cheshire, Connecticut: Graphics Press.

Tufte, Edward (2001). *The Visual Display of Quantitative Information, 2nd Edition.* Cheshire, Connecticut: Graphics Press.

Tufte, Edward (2006). *Beautiful Evidence*. Cheshire, Connecticut: Graphics Press.

Tversky, Amos (2004). *Preference, Belief, and Similarity: Selected Writings by Amos Tversky*. edited by Eldar Shafir. Cambridge, Massachusetts: A Bradford Book of the MIT Press.

Tversky, Amos, and Kahneman, Daniel (1974). Judgment under Uncertainty: Heuristics and Biases, *Science*, New Series, Vol. 185, No. 4157

United States Environmental Protection Agency (1998). *Seven Cardinal Rules of Risk Communication*. Retrieved from at https://www.epa.gov/

US Department of the Army (1984). *Effective Communication, CM-02-RC with changes*. US Army Sergeants Major Academy, Fort Bliss Texas

US MIL-STD-721C (1981). *Military Standard: Definitions of Terms for Reliability and Maintainability*. Department of Defense, Washington, DC.

Vallero, D. A. (2007). *Biomedical Ethics for Engineers: Ethics and Decision Making in Biomedical and Biosystem Engineering*. Burlington, Massachusetts: Elsevier/Academic Press.

Vallero, D. A., & Vesilind, P. A. (2007). *Socially Responsible Engineering: Justice in Risk Management*. Hoboken, N.J.: John Wiley.

Volz KG, and Gigerenzer G. (2012). Cognitive Processes in Decisions Under Risk are not the Same as in Decisions Under Uncertainty. *Frontiers in Neuroscience*. 2012;6:105.

Wilbur, David. Notes from the 2017 Asset Management Ecosystem. Go2Learn. March 27 to 29, 2017. Henderson, NV.

INDEX

A

abbreviations, 102

academics, 60, 63, 119

accountants, 64

accuracy, 31, 128

achievements, 65, 115

action, 24-26, 28-29, 33, 40-41,
    56-57, 84, 93, 96, 100, 123,
    125, 160, 164, 170, 187, 190

adage, 69, 86, 112, 120

Adams,Tim, 169, 184-191

adaptive management, 9, 182

additive, 10

administration, 176

advisor, 28

advocacy, 112, 164

aerospace, 2

affiliation, 65, 159

agency, 9, 63, 118, 131, 168

agenda, 61, 97

airplanes, 145

algorithm, 42

alignment, 33, 136, 190

allocation, 25-27, 30, 33, 56,
    85, 95, 112, 128, 141, 157

alternatives, 24, 60, 170, 172-
    173

American Society for Quality

(ASQ), 64

American Society for Quality
    Reliability Division (ASQ-
    RD), 181

analogy, 125, 189

analysts, 62, 80, 119

analytic, 27, 69, 82, 116

animations, 46

Anscombe's quartet, 35-36, 43

apostle, Paul the , 55

appreciation, 122, 130, 132,
    136

Aristotle, 55, 117

Arrow, Kenneth, 182-183

artists, 48, 80

assessment, 20, 114-116, 123,
    151, 154, 156

asset, 3, 11, 20, 79, 190

at RISK (@RISK), 18, 68

attorney, 18, 64, 93, 99

Aubie, 23-24, 126-127

Auburn University, 23, 126

audible, 91

audience, 3, 17, 38, 49-51, 54,
    56, 65, 70, 72, 75, 78, 85,
    95-96, 103, 110-112, 114-
    118, 120-122, 124, 126,
    128, 130, 132, 134-137,

154, 159-160, 163-165,
167, 169-170, 172-174,
188-191
audience response system, 17
auditory, 120-121
availability, 2-3, 129, 141, 144-
145, 147
avoidance, 111
awareness, 111, 122

41, 52, 74-75
brainstorming, 45, 143
breast cancer, 18, 30
Briggs, Katherine Cook, 113-
116
buffoonery, 55
bureaucracy, 131
business case evaluation (BCE),
171, 178, 190

B

Baggett, Bart, 118, 188
bar chart, 40-41, 43, 45, 52
baseball, 18
Bayesian, 22-23
Bazovsky, Igor, 181
behavioral, 19, 23, 105, 127,
129
beliefs, 128, 168
Belushi, John, 27
Bentham, Jeremy, 56
Bernstein, Peter, 181
bias, 19, 61, 91, 126, 128-130,
132-133, 141, 147, 149,
152, 159, 188, 190
BigPicture, 68
binomial distribution, 153
biological, 11
biomedical, 60, 186
black swans, 141
Blackwell, Joel, 188
Booher, Dianna, 102, 172-173,
188, 190
box and whisker diagram, 40-

C

calculus, 19
camera, 70, 103
canon, of ethics, , 58, 63, 186
career, 12, 27, 97, 100, 131
carousing, 56
cartography, 39
cascading, 101, 155
cash cows, 83
catastrophic, 10, 13
categorization, 106
causation, 82
Centers for Disease Control
(CDC), 167
Certified Reliability Engineer
(CRE), 64
chartjunk, 46-47, 52, 80, 103
checklist, 42, 171, 185
choices, 148, 184
cholera, 39
choleric, 113, 116
Cirisis and Emergency Risk
Communication (CERC),
167-169

clarity, 7, 25, 36, 39, 52, 54-55, 135, 173
Clark, Ruth Colvin, 48, 185
climate change, 54
cognition, 63, 127
collaboration, 50, 184, 187, 189
colleagues, 96, 101, 135
color, 46, 49-52, 71, 74, 78, 80, 102-105, 109, 177-178, 185
colorblind, 49
commissioner, 60, 62, 92, 166
commitment, 70, 150, 170
communication plan, 111, 155, 158, 164-167, 171
communicator, 22, 28, 53-54, 70, 75, 77-78, 84-85, 89, 95, 160, 164
compassion, 170
competence, 160
competition, 176
complexity, 27, 38, 79, 81, 130, 132, 136, 166, 168, 183
compliance, 116, 123
component, 9, 25, 78-79, 92, 112, 151-153, 164, 167, 172, 187-188, 190
conciseness, 176
concrete, 12-13, 144-145
confidence, 22, 51, 59, 80, 141
consensus, 15, 30, 130, 134, 152, 163
consequence, 5, 7, 28, 45-46, 56-57, 59, 64, 83-84, 91, 98-99, 125, 132, 139-142, 147-148, 151-152, 155-156, 163, 186

consequence of failure, 46
consequential ethics, 54, 57-58
consequentialism, 56, 65
consultant, 27, 77, 80, 174, 176-177
continual improvement, 127
controversy, 124, 185
corporations, 130
correlation, 36, 43, 45, 82, 117
Council of Supporting Organizations (COSO), 5, 184
countermeasure, 169
Cox, Anthony, 185
credentials, 159
credibility, 50, 76, 84, 97, 101, 103, 159-161
crisis, 63, 142-143, 152, 167-169, 189-190
criteria, 92, 130, 133-134, 147
critical need for definitions, 15
critically, 8, 26-27, 29
customer, 14, 41, 131, 162, 175

D

dangers, 151
data maps, 39
Dawes, Robyn, 133, 183
decisions, 15, 19, 23-24, 26-27, 30, 33, 54, 56, 69, 71, 80, 104, 112, 122-123, 125-126, 128-131, 134-136, 141, 146-149, 153, 163-164, 168, 178, 183-184

DecisionTools Suite, 68, 186
defect, 13, 42-43
deficiency, 7
delivery, 69, 86, 90, 171, 188
denial, 116, 169
denominator effect, 149-150
deontological, 54, 57-58, 65,
    186
dependability, 2-3, 104, 144-
    145
deterministically, 76
deviation, 6, 14, 155, 187
Dialogue Decison Process
    (DDP), 136
dimensions, 37, 48, 80-81, 123
disaster, 142, 154
DISC, 113, 115-116, 119, 122-
    123
disclosure, 57
discrete distributions, 153
discussions, 8, 181, 187, 189
disease, 167
dispersion, 84, 187
distraction, 47, 49-52, 76, 84
distributions, 43, 73, 84-85,
    153
diversity, 156
Division of Environmental
    Quality (DEQ), 61, 92-93
dollars, 45, 83
downtime, 3
drunkenness, 56
Duke Unviersity, 63
duty, 38, 57-59, 62-65, 93, 186
dyslexia, 185

E

earthquakes, 142
economics, 40, 56
economist, 23, 43
editions, 114
editor, 32, 98-99, 183, 188
educational, 48, 65, 190
effectiveness, 95, 106, 112, 125,
    186
efficiency, 36, 106, 131-132
elected officials, 94-95, 100-
    101, 109, 131, 167, 188
email, 99, 191
emergency, 12, 141, 167-169,
    189-190
emotion, 104, 116, 147, 151
empathy, 158
employee, 7, 41
employers, 58
enemy, 145
enforceable, 59-60, 63, 65
engineer, 13-14, 58, 63-65, 84,
    91-93, 97, 109, 159, 182,
    186, 190
entertainment, 96
environmental, 2, 9, 60, 92-93,
    113, 144, 154, 164, 168,
    182
Environmental Management
    Commission, 60-62, 92
epidemic, 39
equations, 51
equipment, 20, 70, 128
errors, 77, 130, 182, 188
ethic, 38, 53-66, 92-93, 131,

160, 186
Evolver, software by Palisade, 68
expertise, 36, 62, 84
experts, 19, 38, 49, 60, 64, 77, 98, 100, 102, 129, 132-134, 172
explanation, 1, 27, 37, 77-78, 155, 157, 178
exponential, 132
exposures, 145, 151
expression, 5, 7-8, 36, 51, 77
extraversion, 114
eye movements, 120-122

F

facilities, 70, 142, 154
failure, 3, 5, 10, 13, 15, 20-22, 45-46, 65, 73, 83-84, 128, 144-146, 150, 152, 175-176, 181
Failure Modes and Effects Analysis (FMEA), 176
fairness, 59
farce, 80
Federal Emergency Management Agency (FEMA), 168-169, 190
feedback, 50, 106-107
feelings, 120-121
financier, 104
fishbone diagram, 42, 44-45
five stages of grief, 169
flattery, 55
flawed, 84

flow chart, 72, 80
forecaster, 73-74
forefathers, 130
forensic, 99
fornication, 56
frameworks, 170, 184
Franklin, Benjamin, 161
frequentist, 22-23
Freud, Sigmond, 113
Fried, Robert, 131-132, 188

G

Galatians, 55
Galileo, 37
geographic information system (GIS), 68
Gestalt method, 117-119
Gigerenzer, Gerd, 18-19, 30, 181-182
graphology, 117-118
growth-share matrix, 83

H

handwriting analysis, 118-119, 188
hatter, John Thompson, 161-162
hazards, 104
heat maps, 45-46, 83, 185
heuristics, 126, 128-130, 149, 188

hierarchy, 106-107, 110
highlighters, 104
hindsight, 72
Hippocrates, 113
histogram, 40-43, 45, 52, 74,
    84-86, 186-187
hold paramount, 58
honesty, 38
humanity, 97
humans, 9, 18-19, 23, 28, 126,
    141, 147, 163, 182, 186
hurricane, 142, 148
hypothesis, 163

I

ideas, 36, 62, 89
illness, 9, 139
illogical, 147, 150
imagination, 108, 154
incompetence, 160
incremental, 56, 169
inference, 22
infrastructure, 11, 54, 70, 164
innumeracy, 18
insight, 22, 51, 70, 122, 168,
    181, 183, 188
instinct, 91, 183
instrumentation diagram, 48
integer, 43, 46
integrity, 14, 26, 37, 47, 52, 64
intention, 26, 167
internet, 80, 119, 164, 171, 188
interpretation, 51, 98, 101, 109,
    117

interval, 2, 22, 72
introversion, 114
intuition, 114, 127-128
intuitive, 7, 18, 23, 33, 84-85,
    115, 127, 165, 168
investment, 146
investors, 28
Ishikawa diagram, 42, 44
ISO 31000, 5-8, 15, 73, 79
ISO 55000, 79
jargon, 67, 160

J

journalism, 99
judgement, 19, 83, 127, 129,
    149, 173
Jung, Carl Gustav, 113-118,
    188
Juran, Joseph, 185
justice, 64

K

Kahneman, Daniel, 19, 23, 28,
    129-130, 135, 146-149,
    182-184, 188-189
Kant, Immanuel, 57
Keeney, Ralph, 183
kinesthetic, 120-121
Knight, Frank, 4, 7-8, 23, 181-
    182, 188
Kubler-Ross, Margaret, 169

L

Lambert, J.H., 39, 41-42
laws, 59, 63-64
lawsuit, 62
layout, 102, 177-178
leader, 19, 39, 106-107, 130-131
leadership, 9, 131
Leavitt, Harold J., 106, 188
legislation, 57, 92
legislative, 92, 141
liability, 13-14
licensure, 63
likelihood, 5, 9, 20-22, 45-46, 83-84, 128, 139-142, 145-148, 152, 155-156
likelihood of failure, 5, 20-22, 46, 83-84, 128, 145-146
Likert scales, 46, 188
line and time series graph, 39, 41, 52
linear, 36, 62, 133
lobbyist, 100
logic, 104, 127, 134, 163
losses, 147-148

M

machinery, 9
Mahalanobis distance, 62
Maier, Pauline, 161-162
maintainability, 2, 145, 181, 190

maintenance, 3, 65, 128, 144
malpractice, 186
management, 4-5, 9, 20, 41, 60-61, 79-80, 92, 114, 119, 132, 142, 153, 166, 168, 176-177, 181-182, 184-185, 189-190
managers, 131, 166, 184
manipulation, 38, 48, 62, 95, 112, 165-166
manufacturers, 70
map, 39, 45, 172
March, James, 130, 188
Marston, William Moulton, 115-117, 122-123, 188
mascot, 23, 126
mathematics, 22, 91
matrices, 83-84, 86, 185
matrix, 83
MBTI, 114-116, 119
measurement, 8, 21, 37, 73, 83, 130
media, 46, 94-95, 97-101, 108-109, 155, 165, 167, 170, 176-177, 185, 188
Meehl, Paul, 133, 183
Meet the Fockers, 121
meetings, 26, 92, 94, 191
messenger, 27-28, 33, 54, 174
Mill, John Stuart, 56
misrepresentation, 31, 62
mission, 3, 63, 99, 182
mitigation, 83, 143, 184
Mitroff, Ian, 142, 189
modes, 176, 188
monetary, 37, 72, 83-84

Monte Carlo analysis, 68, 149, 152-153
multivariate, 36
Myers, Isabel Briggs, 113-115, 187

N

NASA, 182, 189
NASCAR, 3
naturalist, 104
negotiation, 115, 173-175
NeuralTools, 68
neuro-linguistic programming, 113, 119, 188
newspaper, 98-99
noise, 47, 49-50, 89-92, 94, 96-98, 100-111, 123, 125, 132, 137, 155, 159-160, 172, 175, 187, 191
numerator, 149
numerical, 37, 129
nutrients, 60

O

O'Connor, Patrick, 181
Obamacare, 56
objectives, 5-8, 14, 109, 134, 146, 166-167
objectivity, 38, 64
observations, 20, 42, 49, 72
olfactory, 91

operations, 41, 128, 144, 176
opportunities, 83, 100, 103
optimization, 68, 83
organization, 5, 7, 22, 26, 44, 60-61, 63, 65, 73, 96, 106-107, 114, 130-132, 134-135, 158, 163, 171, 190
orientation, 59, 71-72
outcome, 39, 56, 73-74, 95, 148, 157, 186
outliers, 36, 62
ownership, 141

P

pacific weather patterns, 31
Palisade, 18, 68-69, 186
parameter, 73, 77-78
Pareto diagram, 42-44
passion, 56, 104, 168, 188
pastels, 49, 71, 103
Paul, the apostle, 55
peers, 48, 106, 130
percentage, 5, 20, 22, 30-32, 76, 123, 144
perception, 46, 64, 73, 114, 141, 146, 148, 189
performance, 9, 23, 58, 130-131, 144, 170
persuasion, 48, 95, 101, 112, 164-165, 184
philosopher, 55, 117
philosophy, 39
photographs, 71
physicians, 58, 64-65

picture, 46, 69-71, 86, 89, 97, 162

pie charts, 40-41, 46, 52, 85-86

Planning Fallacy, 135

Playfair, William, 39-42

Poisson distribution, 153

policies, 65

politicial action committee (PAC), 100

politician, 101, 148

politicization of science, 60

portfolio, 83

practitioners, 49, 54, 83

precipitation, 17-18, 182

precision, 36, 80

PrecisionTree, 18, 68

prediction, 19, 31, 73, 129, 133-134

preference, 69, 84, 115, 173, 183

prevention-based, 144

principle, 37, 52, 58, 186

prioritization, 141, 163

probabilistic, 18, 73-75, 149, 152-153, 173

probability, 2-3, 8, 13-14, 18-23, 32, 85-86, 129, 140, 147-150, 153, 173, 182-183

procedures, 42, 65, 164

products, 41, 68, 128, 131

professionalism, 51

profile, 113, 118, 165, 188

profitability, 131

profitable, 149

Project Management Institute, 5

projections, 74

proof, 77, 117

proportional, 37, 59

Prospect Theory, 147-149, 189

psychologist, 19, 23, 105, 113, 115-116, 118-119, 127, 130

psychology, 19, 63, 114, 129, 134, 146

public speaking, 94-97, 100, 109, 180

Q

quadrant, 113-114, 122-123, 147

qualifications, 65

quantitative, 2, 36, 74, 83, 152

quartile, 74

R

Raiffa, Howard, 130, 134, 183

rain-outs, 18

rainfall, 182

ramification, 20, 73, 127

randomness, 19

rapport, 119, 121-122, 159

ratio, 90, 109, 149, 187

realists, 96-97

reality, 10, 38, 56, 58, 68, 132

receiver, 78, 89-90, 101, 109-112, 119, 137, 150, 158-160, 162, 164, 169, 172, 174, 178, 187

recognition, 7, 123, 135
redundancy, 175
regression, 36, 62
regulations, 59
regulatory, 92, 128, 141
relationship, 40, 43-45, 70, 72-74, 79, 81-83, 116, 121-122, 127, 146, 151, 157, 189
reliable, 3, 10, 13, 152, 170
Reliasoft, 68
repairable systems, 3
reporter, 98-99, 177
reputational, 142
researchers, 27, 31-32, 60, 62-63, 149
resilience, 15, 33, 91, 96, 111-112, 182
resilient, 9-10, 15, 140, 143-144, 146, 156
risk register, 6, 154
risk-based, 144-145, 156, 163
RiskOptimizer, 68
roads, 12
robust, 67, 133

S

safety, 58, 64, 154, 186
salesperson, 159
Savage, Leonard "Jimmie", 23, 182-183
Savage, Sam, 182
scatter diagram, 42-43, 45, 81-82, 86

scenario, 14, 129, 135, 152, 156, 190
Schoemaker, Paul, 132, 134, 183
scientific, 32, 39, 117, 119, 156, 163, 190
scientists, 31, 60, 63, 97, 109, 127, 150
sea level rise, 141-142
sender, 89-90, 100, 110-112, 118-119, 121, 159, 164, 169, 178
Sharpe, Adam, 85, 166, 171, 187, 190
shootings, police, 30
showmanship, 97
signal-to-noise, 90, 109, 187
Simon, Herbert, 130, 134, 188
Six Sigma, 83
Slovic, Paul, 129, 146, 148-151, 182-184, 188-189
social media, 98, 108, 155, 170
Society of Maintenance and Relaibility Professionals (SMRP), 65
Socrates, 55
Sowers, Roy G., 165, 187
specialist, 38, 49, 72, 109, 127, 166
speeches, 96, 102
Spetzler, Carl, 135-136, 183
spirituality, 105
stakeholder, 93, 96, 130, 134, 141, 156, 183
standards, 2, 5, 8, 14, 58, 60, 63-64, 144, 181

Stanford University, 24-25, 170, 183
Stanovich, Keith, 127, 183
State Bureau of Investigation (SBI), 99
statistician, 23, 35
statistics, 21-22, 31-32, 36-37, 40, 51-52, 68, 92, 135
StatTools, 68
statutes, 64
stereotypes, 165
stormwater, 92
success, 9, 69, 149-150, 153
superiors, 48, 107, 109
survival, 126, 156
sustainability, 9, 182
symbols, 36, 51
System 1, 23-25, 27, 33, 127-128, 168, 183
System 2, 23-25, 27, 33, 127-128, 168, 183
systems, 3, 9-11, 73, 140, 144-146, 156, 182

T

Taleb, Nassim Nicholas, 183
technique, 43, 45, 76, 94, 99, 169, 188, 191
technology, 61, 96
templates, 178
terrorism, 147
terrorist, 24
testimony, 99
Tetlock, Phillip, 134

thermographic, 70
threat, 96, 159-161
timber piles, 12-13
toothless tiger, 24, 126
tornado diagram, 69, 76-79, 86, 147
Total Quality Management (TQM), 41
trainer, 103, 115
Trait Stroke method, 118-119
transportation, 13-15
Treadway Commission, 5
Trump, Donald, 108
trustee, 100
truth, 5, 31, 36, 38, 52-54, 65, 75, 119, 145, 166
tsunami, 140
Tufte, Edward, 36-37, 39-40, 47, 52, 159-160, 184-185, 190
Tukey, John, 39-40
Tversky, Amos, 19, 28, 129-130, 135, 146-149, 182-184, 188-189

U

uncertainty, 2, 4-10, 23, 27, 65, 68, 73-75, 77, 81, 126, 129-130, 132, 135-136, 145-147, 149, 151, 158, 160, 168-169, 173, 178, 181, 183
United States Environmental Protection Agency

(USEPA), 9, 168, 170, 190
unknown-unknowns, 145
unlicensed, 60
unreliable, 152
uptime, 3
urn, 149-150

workflow, 44
workforce training and
    development, 48
written image, 102

V

Vallero, Dan, 60, 89, 91, 185-
    187, 189
variables, 42-43, 45, 76, 82
variation, 37, 47, 51, 76-77, 82,
    187
verdict, 99
Vining, Geoffrey, 62
virtue, 54-56, 65, 186
visual image, 76, 80, 98, 102,
    105, 110, 120
visuals, 35, 37, 48, 85, 121,
    172, 185
vocal image, 102, 105, 108,
    110, 188
voters, 148

W

water transmission system, 175-
    176
wealth, 50, 104
weightings, 148
welfare, 58, 64, 186
whistleblower, 62

75789213R00127

Made in the USA
Columbia, SC
24 August 2017